国外优秀数学著作
原版系列

黎曼–希尔伯特问题与量子场论
——可积重正化、戴森–施温格方程

Riemann-Hilbert Problem and Quantum Field Theory
—Integrable Renormalization, Dyson-Schwinger Equations

[伊朗] 阿里·索贾–法尔德（Ali Shojaei-Fard） 著

（英文）

哈尔滨工业大学出版社
HARBIN INSTITUTE OF TECHNOLOGY PRESS

黑版贸审字 08–2019–180 号

Copyright © 2012 by the author and LAP LAMBERT Academic Publishing GmbH & Co. KG and licensors
All rights reserved. Saarbrücken 2012

图书在版编目(CIP)数据

黎曼-希尔伯特问题与量子场论:可积重正化、戴森-施温格方程 = Riemann–Hilbert Problem and Quantum Field Theory:Integrable Renormalization,Dyson–Schwinger Equations:英文/(伊朗)阿里·索贾-法尔德(Ali Shojaei–Fard)著.—哈尔滨:哈尔滨工业大学出版社,2022.8
ISBN 978-7-5767-0278-1

Ⅰ.①黎… Ⅱ.①阿… Ⅲ.①希尔伯特问题-研究-英文②量子场论-研究-英文 Ⅳ.①O177.1②O413.3

中国版本图书馆 CIP 数据核字(2022)第121739号

LIMAN-XIERBOTE WENTI YU LIANGZI CHANGLUN:KEJI CHONGZHENGHUA,DAISEN–SHIWENGE FANGCHENG

策划编辑	刘培杰　杜莹雪	
责任编辑	张永芹　邵长玲	
封面设计	孙茵艾	
出版发行	哈尔滨工业大学出版社	
社　　址	哈尔滨市南岗区复华四道街10号　邮编150006	
传　　真	0451–86414749	
网　　址	http://hitpress.hit.edu.cn	
印　　刷	黑龙江艺德印刷有限责任公司	
开　　本	886 mm×1 230 mm　1/32　印张6.25　字数200千字	
版　　次	2022年8月第1版　2022年8月第1次印刷	
书　　号	ISBN 978-7-5767-0278-1	
定　　价	38.00元	

(如因印装质量问题影响阅读,我社负责调换)

Preface

This monograph focuses on some recent applications of the Hopf algebraic perturbative renormalization in the study of Quantum Field Theory. In the first part, we introduce integrable renormalization formalism as a method of studying integrable systems underlying the Hopf algebra of Feynman diagrams. Besides we consider an alternative approach to quantum integrable systems which is strongly connected with the infinite dimensional complex Lie group of the renormalization Hopf algebra. In the second part, we consider the process which allows us to extend the Connes-Marcolli universal approach to the study of non-perturbative Quantum Field Theory. For this purpose, at the first step, we work on a new combinatorial representation of the universal Hopf algebra of renormalization H_U in the context of Hall rooted trees which can be lifted onto the level of (universal) counterterms. We study this shuffle type universal Hopf algebra in a combinatorial framework. At the second step, we associate a categorical construction to each Dyson-Schwinger equation DSE which can encode this class of equations with respect to a special family of connections. At the third step, on the basis of the new mentioned picture of H_U, we prove the universality of the category of the equi-singular flat vector bundles at the level of Dyson-Schwinger equations. At the fourth step, applying the mentioned categorical investigation, we determine a new family of Picard-Fuchs type equations (in the context of the cohomological Milnor fibration) related to each equation DSE. In addition, we relate a category of Feynman motivic sheaves (as a subcategory of the Arapura category) to each equation DSE.

MSC 2000: 05C05; 22E65; 37K25; 81Q30; 81T15; 81T70
PACS: 02.10.De, 11.10.Gh, 11.10.Hi, 2.30.Ik, 02.10.Ox, 02.20.Sv, 02.20.Tw, 11.10.-z
Keywords: Combinatorial Hopf Algebras; Combinatorial Dyson-Schwinger Equations; Connes-Kreimer-Marcolli Perturbative Renormalization; Hall Rooted Trees; Motivic Feynman Rules; Picard-Fuchs Equations; Quantum Integrable Systems; Universal Hopf Algebra of Renormalization
Email: shojaeifa@yahoo.com

Acknowledgments

The author would like to thank Prof. Dr. Matilde Marcolli for her helpful ideas, advices and great scientific supports. In addition, the author would like to thank from Hausdorff Research Institute for Mathematics (HIM) for the program Geometry-Physics, Max Planck Institute for Mathematics (MPIM) as a member of IMPRS program, Erwin Schrodinger International Institute for Mathematical Physics (ESI) for the program Number Theory and Physics because of their supports during the work on this monograph.

Contents

1 **Introduction**   5

2 **Hopf Algebra Structures on Combinatorial Objects**   11
   2.1 Elementary properties 11
   2.2 Combinatorial Hopf algebras 16
     2.2.1 Connes-Kreimer Hopf algebra 17
     2.2.2 Rooted trees and (quasi-)symmetric functions 20
     2.2.3 Incidence Hopf algebras 23

3 **Perturbative Renormalization**   27
   3.1 Insertion operator: From a pre-Lie algebra structure on Feynman diagrams to a Hopf algebra 28
   3.2 Hopf algebraic formalism 35

4 **Integrable Renormalization**   43
   4.1 Integrable systems: from finite dimension (geometric approach) to infinite dimension (algebraic approach) 44
   4.2 Rota-Baxter type algebras 47
   4.3 Theory of quantum integrable systems 49
   4.4 Integrable systems on the basis of the renormalization group 56
   4.5 Baditoiu-Rosenberg framework: The continuation of the standard process 62
   4.6 Fixed point equations 63

5 **Connes-Marcolli Theory**   68
   5.1 Geometric nature of counterterms: Category of flat equi-singular connections . . . 68
   5.2 The construction of a universal Tannakian category 71

6 **Universal Hopf Algebra of Renormalization**   75
   6.1 Shuffle type representation 75
   6.2 Rooted tree type representation 78
   6.3 Universal counterterms 91

7 **Combinatorial Dyson-Schwinger Equations and Connes-Marcolli Universal Treatment**   96
   7.1 Quantum motions in terms of the renormalization Hopf algebra 97

7.2 Universal Hopf algebra of renormalization and factorization problem 102
7.3 DSEs in a categorical framework 108

8 **From Combinatorial Dyson-Schwinger Equations to the Category of Feynman Motivic Sheaves** **113**
8.1 DSEs in the context of the theory of motives 113

9 **Conclusion** **119**

编辑手记 129

Chapter 1

Introduction

... " *You see, one thing is that I can live with doubt and uncertainty and not knowing. I think it's much more interesting to live not knowing than to have the answers that might be wrong* ", Richard Feynman ...

Quantum Field Theory (QFT) is the most important profound formulated manifestation in modern physics for the description of occurrences at the smallest length scales with highest energies. In fact, this mysterious theoretic hypothesis is the fundamental result of merging the two crucial achievements in physics namely, Quantum Mechanics and Special Relativity such that its essential target can be summarized in finding an unified interpretation from interactions between elementary particles. There are several formalisms which are aimed for developing QFT where Wightman's axiomatic configuration (i.e. constructive QFT) together with Haag's algebraic formulation in terms of von Neumann algebras are well-known influenced efforts in this story [48, 91, 111]. Perturbation approach is also another successful and useful point of view to QFT. It is based on perturbative expansions related to Feynman graphs where in these expansions one can find some ill-defined iterated Feynman integrals such that they should be removed in a physical sensible procedure. This problem can be considered in the context of a well-known analytic algorithm namely, renormalization theory. [22, 73, 81, 91]

The apparatus of renormalization developed in the perturbation theory led to attractive success in quantum field theory where it provides in fact an analytic meaning. But Kreimer could find out the appearance of a combinatorial nature inside of this analytic method and he showed that one can explain the renormalization procedure (as a recursive formalism for the elimination of (sub-)divergences from diagrams) based on one particular Hopf algebra structure H_{FG} on Feynman diagrams. The discovery of this smooth mathematical construction encapsulated in the renormalization process covered the lack of a modern practical mechanism for the principally description of this technique and also, it introduced a new rich relationship between this essential physical method and modern mathematics. It should be mentioned that although the Bogoliubov recursion performs renormalization without using any Hopf algebra structure but when we go to the higher loop orders, the advantages of this Hopf algebraic reconstruction in computations will be observed decisively. [8, 9, 31, 32, 52, 62, 63, 65, 109, 110]

There are different approaches to renormalization. For instance, in the Bogoliubov method, the renormalization is done without regularization and counterterms but in the Bogoliubov-Parasiuk-Hepp-Zimmermann (BPHZ) method, we work on dimensional regularization (as the regularization scheme) and minimal subtraction (as the renormalization map). Originally, the regularization can parametrize ultraviolet divergences appearing in amplitudes to reduce them formally finite together with a special subtraction of ill-defined expressions (associated with physical principles). But this procedure determines some non-physical parameters and indeed, it changes the nature of Feynman rules (identified by the given physical theory) to algebra morphisms from the renormalization Hopf algebra H_{FG} to the commutative algebra A_{dr} of Laurent series with finite pole part in dimensional regularization. It is obviously seen that this commutative algebra is characterized with the given regularization method and it means that by changing the regularization scheme, we should work on its associated algebra. In general, interrelationship between Feynman diagrams and Feynman integrals can be described by the Feynman rules of the theory and Kreimer could interpret this rules on the basis of characters of H_{FG}. [52, 57, 58, 59]

Soon thereafter, Connes and Kreimer introduced a new practical reformulation for the BPHZ perturbative renormalization. They could associate an infinite dimensional complex Lie group $G(\mathbb{C})$ to each renormalizable physical theory Φ and then they investigated that this BPHZ scheme is in fact an example of a general mathematical procedure namely, the extraction of finite values on the basis of the Riemann-Hilbert problem in the sense that one can calculate some important physical information for instance counterterms and related renormalized values with applying the Birkhoff decomposition on elements of this Lie group. In other words, according to their programme, based on the regularization scheme, the bare unrenormalizaed theory produces a meromorphic loop γ_μ on $C = \partial \Delta$ with values in the space of characters (i.e. the Lie group $G(\mathbb{C})$) where Δ is an infinitesimal disk centered at $z = 0$. For each $z \in C$, $\gamma_\mu(z) = \phi^z$ is called dimensionally regularized Feynman rules character which means that the special character ϕ sends each arbitrary Feynman diagram Γ to a generally divergent iterated Feynman integral but the regularized version ϕ^z of this character can map each Feynman diagram Γ to a Laurent series $U_\mu^z(\Gamma)$ in z with finite pole part (i.e. regularized unrenormalized version of the related Feynman integral). In this machinery, the renormalized theory can be formulated with the evaluation at the integer dimension D of space-time of the holomorphic positive part of the Birkhoff decomposition of γ_μ. As the result, Connes and Kreimer could recompute some important physical information such as counterterms and related renormalized values, renormalization group and the associated infinitesimal generator (i.e. β−function) with respect to components of the Birkhoff decomposition of γ_μ (which is a loop with values in the group of formal diffeomorphisms of the space of coupling constants). [14, 15, 19, 22, 40]

The existence and the uniqueness of the Connes-Kreimer's Birkhoff decomposition are connected with the Rota-Baxter property of the chosen regularization and renormalization couple (i.e. multiplicativity of renormalization). This fact reports a new bridge between the theory of Rota-Baxter type algebras and the Riemann-Hilbert correspondence such that as the consequence, one can expect to study quantum integrable systems in this Hopf algebraic language. [27, 28, 29, 40, 93]

Connes and Marcolli widely improved this mathematical machinery from perturbative renormalization by giving a categorical algebro-geometric dictionary for the analyzing of physical information in the minimal subtraction scheme in dimensional regularization underlying the Riemann-Hilbert correspondence. According to their approach, the dimensional regularization parameter $z \in \Delta$ determines a principal $\mathbb{C}^*$-bundle B over the infinitesimal disk Δ (i.e. $p : \Delta \times \mathbb{C}^* \longrightarrow \Delta$). Letting $B^0 := B - p^{-1}\{0\}$ and $P^0 := B^0 \times G(\mathbb{C})$. Each arbitrary equivalence class $\bar{\omega}$ of flat connections on P^0 denotes a differential equation $\mathbf{D}\gamma = \bar{\omega}$ (i.e. $\mathbf{D} : G(\mathbb{C}) \longrightarrow \Omega^1(\mathfrak{g}(\mathbb{C}))$, $\phi \longmapsto \phi^{-1} d\phi$) such that it has a unique solution. Equi-singularity condition on $\bar{\omega}$ describes the independency of the type of singularity of γ at $z = 0$ from sections of B. It means that for sections σ_1, σ_2 of B, $\sigma_1^*(\gamma)$ and $\sigma_2^*(\gamma)$ have the same singularity at $z = 0$. Connes and Marcolli firstly could reformulate components of the Birkhoff factorization of loops $\gamma_\mu \in Loop(G(\mathbb{C}), \mu)$ based on time-ordered exponential and elements of the Lie algebra $\mathfrak{g}(\mathbb{C})$. Secondly, they found a bijective correspondence between minus parts of this kind of decomposition on elements in $Loop(G(\mathbb{C}), \mu)$ (which determine counterterms) and elements in $\mathfrak{g}(\mathbb{C})$ and finally, they proved that each element of this Lie algebra determines a class of flat equi-singular connections on P^0. As the conclusion, one can see that this family of connections encode geometrically counterterms such that the independency of counterterms from the mass parameter μ is equivalent to the equi-singularity condition on connections.

In addition, they showed that these classes of connections can play the role of objects of a category $\mathcal{E}^\Phi$ such that it can be recovered by the category $\mathcal{R}_{G^*}$ of finite dimensional representations of the affine group scheme G^*. In the next step and in a general configuration, they introduced the universal category of flat equi-singular vector bundles $\mathcal{E}$ such that its universality comes from this interesting notion that for each renormalizable theory Φ, one can put its related category of flat equi-singular connections $\mathcal{E}^\Phi$ as a full subcategory in $\mathcal{E}$. Because of the neutral Tannakian nature of this universal category, one important Hopf algebra can be determined from the procedure. That is universal Hopf algebra of renormalization $H_\mathbb{U}$. By this special Hopf algebra and its associated affine group scheme $\mathbb{U}^*$, renormalization groups and counterterms of renormalizable physical theories have universal and canonical lifts. Connes and Marcolli developed this strong mathematical treatment to the motivic Galois theory. This is the situation that we can observe the application of the theory of motives for describing physical theories underlying Galois groups and the Grothendiecks standard conjectures. In [82, 83] the author considers this natural motivic investigation at the center of the renormalization theory underlying the Picard-Fuchs type regular singularities and Mellin transforms and then interrelationship between this structure and Connes-Marcolli geometric approach (on the basis of flat equisingular connections) is studied. As the consequence, a new motivic framework for the dimensional regularization procedure is introduced. [3, 6, 19, 20, 21, 22, 80, 81, 82, 83]

Recently, we can observe several interesting progresses about the construction of motivic Feynman rules and their properties in the context of the standard conjectures of Grothendieck. Following this procedure, the parametric representation of Feynman integrals applies to relate these integrals to graph hypersurfaces which allow us to formulate motivic Feynman rules (as an algebrogeometric formalism) and then reinterpret dimensional regularization and so the Connes-Kreimer theory in the context of the theory of motives. Marcolli and coauthors could introduce a new approach to Feynman integrals in the language of motives where the dimensional regularization process and the Connes-Kreimer perturbative renormalization would be lifted to a motivic configuration. According to their methodology, parametric representation of Feynman integrals leads to formulate them as integrals on projective spaces which apply to handle divergences as the intersections of the related graph hypersurface and a cycle (as the domain of integration) in a relative homology group. So applying this reformulation allows us to consider Feynman integrals at the level of relative cohomology which can be claimed as the realization of mixed Tate motives. This formalism convinces us to concern a conjectural relation between residues of Feynman integrals and periods of mixed Tate motives. The parametric representation of Feynman integrals is very useful tool to considering the Connes-Kreimer approach at the level of the specializations of Tutte polynomial such as Jones polynomial, chromatic polynomial and Tutte-Grothendieck polynomial. Indeed, Tutte-Grothendieck invariants of graphs apply to characterize a class of graphs that behaves well with respect to deletion-contraction relations. Using the multiplicative property over disjoint union of graphs shows that the Tutte polynomial invariant and the Tutte-Grothendieck invariant determine abstract Feynman rules with values in $\mathbb{C}[x,y]$ and $\mathbb{C}[\alpha,\beta,\gamma,x,y]$. Moreover, this parametric approach together with the Kirchhoff polynomials can handle Feynman rules and Feynman integrals in the language of graph hypersurfaces and with respect to Kirchhoff-Symanzik polynomials such that it leads to consider Feynman integrals as periods of algebraic varieties. [3, 4, 81, 82, 83]

In the context of this view of point, it will be possible to formulate an algebro-geometric Feynman rule as an invariant multiplicative $U(\Gamma) = U(\mathbb{A}^n \backslash \widehat{X}_\Gamma)$ with values in the regularization algebra A. It assigns a polynomial invariant in $\mathbb{Z}[T]$ to the class in $\mathcal{F}$ of each hypersurface complement $\mathbb{A}^n \backslash \widehat{X}_\Gamma$. An algebro-geometric Feynman rule is called motivic, if the Feynman rule U only depends on the class $[\mathbb{A}^n \backslash \widehat{X}_\Gamma]$ of the hypersurface complement in the Grothendieck ring of varieties $K_0(\mathcal{V}_\mathbb{Q})$. Pursuing this program results the universal algebro-geometric Feynman rule which sends each Feynman diagram to the class of hypersurface complement $\mathbb{A}^n \backslash \widehat{X}_\Gamma$ in $\mathcal{F}$.

This motivic configuration can be studied in another direction as an assignment of an Euler characteristic χ_{new} to the graph hypersurface complements $\mathbb{P}^{n-1} \backslash X_\Gamma$ with the multiplicative property. The authors in [3, 4, 6, 81, 82, 83] construct the process which produces a natural modification of χ_{new} of the usual Euler character with the multiplicative property. As the result, using the multiplicative property of the affine hypersurface complements over disjoint unions of graphs, a universal algebro-geometric Feynman rule which takes values in a suitably defined Grothendieck ring $\mathcal{F}$ of immersed conical varieties can be formulated. This universal Feynman rule can be mapped to a motivic Feynman rule with values in the usual Grothendieck ring of varieties. Using this framework shows that the motivic Feynman rules can be presented by

$$\Gamma \longmapsto [\mathbb{A}^n \backslash \widehat{X}_\Gamma] = (\mathbb{L} - 1)[\mathbb{P}^{n-1} \backslash X_\Gamma] \in K_0(\mathcal{V}_\mathbb{Q}) \quad (1.0.1)$$

together with the multiplicative property such that its renormalized version will be $U(\Gamma) = \frac{[\mathbb{A}^n \backslash \widehat{X}_\Gamma]}{\mathbb{L}^n}$ with values in the ring $K_0(\mathcal{V}_\mathbb{Q})[\mathbb{L}^{-1}]$. Applying the homomorphism of rings I_{CSM} which is defined in terms of Chern-Schwartz-MacPherson characteristic classes of singular algebraic varieties, one can produce a new algebro-geometric Feynman rule. [3, 4]

The systematic extension of this Hopf algebraic modeling to different kinds of (local) quantum field theories would be an attractive topic for people and we can observe the improvement of this aspect for example in the reformulation of Quantum Electrodynamics (QED) (i.e. describes the interaction of charged particles such as electrons with photons), Quantum Chromodynamics (QCD) (i.e. describes the strong interaction between quarks and gluons) and quantum (non-)abelian gauge theories. [7, 60, 64, 90, 101, 102, 103, 108]

Furthermore, finding a comprehensive description from nonperturbative theory based on the Riemann-Hilbert problem is also known as an important and interesting challenge in this Hopf algebraic viewpoint. In [19] the authors suggest a procedure by the Birkhoff factorization and the effective couplings of the (renormalized) perturbative theories. This recommended way allows us to considering nonperturbative theory in the context of the Riemann-Hilbert correspondence where we enable to apply the methods of summation of divergent series modulo functions with exponential decrease of a certain order which is called Borel summability. At the end of the day, a new interpretation from nonperturbative quantum field theory under local wild fundamental group will be produced. [6, 19, 20, 21, 22]

On the other hand, there is a specific class of equations in physics for the analytic studying of nonperturbative situations, namely Dyson-Schwinger equations (DSEs). With attention to the combinatorial nature of the Hopf algebra of renormalization (as the guiding structure) and also, with the help of Hochschild cohomology theory, Kreimer introduced a new significant combinatorial version from these equations. Working on classification and also calculating explicit solutions for these equations eventually lead to a constructive achievement for the much better understanding of nonperturbative theory. The recursive nature of this type of equations leads to study the structure of β-functions in QED and QCD under perturbation theory and ordinary differential equations. In addition, with respect to the renormalization Hopf algebra, Kreimer and colleagues could rewrite DSEs in terms of derivatives of the Mellin transforms for primitives where as the consequence, they calculated β-functions from global solutions of these equations. [8, 9, 52, 53, 55, 59, 61, 62, 63, 70, 91, 106, 107, 108]

The study of this kind of equations on different Hopf algebras of rooted trees can help us to improve our knowledge for the identification of combinatorial nature of nonperturbative events and its reason comes back to this essential note that Hopf algebras of renormalizable theories are representable with a well-known Hopf algebra on rooted trees, namely Connes-Kreimer Hopf algebra. Some interesting results about the study of DSEs on rooted trees are collected in [35, 36, 43, 44].

In short, combining the perturbation theory, the combinatorics of renormalization, the geometry of dimensional regularization, the Connes-Marcolli categorification method and combinatorial Dyson-Schwinger equations leads people to search about a perfect conceptual understanding of quantum field theory in the context of advanced methods of mathematics.

Now let us explain the structure of the present monograph. In this research work, we are going to consider the applications of this Hopf algebraic formalism in the study of some essential problems in theoretical physics namely, integrable systems and nonperturbative theory. In the first purpose, we familiar with a new family of integrable systems which depends upon the perturbative renormalization and the theory of noncommutative differential forms. It will be shown that how Hopf algebraic renormalization can encode new geometric information about perturbative quantum field theory. In the second purpose, we introduce a new categorical approach in the study of Dyson-Schwinger equations with respect to the Riemann-Hilbert correspondence. It will be discussed that how this formalism can help us to interpret these equations in the context of the theory of motives. We start with an overview about Hopf algebras and a particular family of them namely, combinatorial Hopf algebras. In section three, we familiar with the essential details of the Connes-Kreimer theory. The section four contains our approach to integrable systems in renormalizable quantum field theories. After a brief overview of the Connes-Kreimer formalism in section five, we study accurately the structure of the universal Hopf algebra of renormalization such that in section six we give a new Hall rooted tree representation from this Hopf algebra at different levels (i.e. Lie group and Lie algebra levels). This new reformulation allows us to consider interesting relations between this particular Hopf algebra and other combinatorial Hopf algebras and in addition, it will be applied to recalculate physical information such as counterterms in a universal treatment. The section seven is the place to study DSEs where we will associate a category to each equation. This categorification in the basis of some geometric objects leads us to introduce a new approach in the study of DSEs at the universal level with respect to the Riemann-Hilbert problem. On the other hand, because of the motivic nature of the category $\mathcal{E}$, we will enable to produce some new motivic information hidden inside of DSEs which reports about nonperturbative phenomena. In section eight, following the recent observations ([3, 4]), we are going to modify the motivic renormalization framework ([82, 83]) to study DSEs at the level of the theory of motives. For this purpose, we apply the categorical configuration (connected with these equations) in the context of the Riemann-Hilbert correspondence introduced in [99] and Marcolli's framework [83] to produce a new family of Picard-Fuchs equations which reports some geometric information hidden inside of nonperturbative theory. In addition, we focus on the Aluffi-Marcolli approach to motivic Feynman rules to produce a category of Feynman motivic sheaves which is related to primitive components of unique solution of a given DSE.

Chapter 2

Hopf Algebra Structures on Combinatorial Objects

The concept of Hopf algebra was introduced based on the work of Hopf on manifolds and then its widely applications in topology, representation theory, combinatorics, quantum groups and non-commutative geometry displayed the power of this structure in different branches of mathematics [1, 50, 96]. Indeed, it is important to know that Hopf algebras provide generalizations for group theory and Lie theory. As a well-known example, it can be seen that the dual of the universal enveloping algebra of a simple Lie algebra determines a Hopf algebra. Moreover, this powerful mathematical construction can provide a new opportunity to find useful interrelationships between the pure world of mathematics and some complicate techniques in modern physics such as perturbative renormalization. Additionally, with the help of a special class of Hopf algebras namely, quantum groups, one can observe the developments of this theory in mathematical physics and theoretical physics [49, 51, 79, 86, 109].

Since Hopf algebras play a skeleton key for this work, therefore it is essential to have enough information about them. With attention to our future requirements, in this chapter we familiar with the concept of Hopf algebra and then we will have a short overview from its basic properties.

2.1 Elementary properties

Let $\mathbb{K}$ be a field with characteristic zero. A $\mathbb{K}$-vector space A together with an associative bilinear map $m_A : A \otimes A \to A$ and a unit $\mathbf{1}$ is called *unital algebra* such that its associativity and its unit are expressed respectively by the following commutative diagrams:

$$\begin{array}{ccc} A \otimes A \otimes A & \xrightarrow{m_A \otimes id_A} & A \otimes A \\ {\scriptstyle id_A \otimes m_A} \downarrow & & \downarrow {\scriptstyle m_A} \\ A \otimes A & \xrightarrow{m_A} & A \end{array} \qquad (2.1.1)$$

$$\mathbb{K} \otimes A \xrightarrow{\mu \otimes id_A} A \otimes A \xleftarrow{id_A \otimes \mu} A \otimes \mathbb{K} \qquad (2.1.2)$$
$$\searrow_{\sim} \quad \downarrow_{m_A} \quad \swarrow_{\sim}$$
$$A$$

where the map $\mu : \mathbb{K} \longrightarrow A$ is defined by $\mu(\lambda) = \lambda \mathbf{1}$. This algebra is *commutative*, if $m \circ \tau = m$ such that $\tau : A \otimes A \to A \otimes A$ is the *flip map* defined by

$$\tau(a \otimes b) = b \otimes a. \tag{2.1.3}$$

With reversing all arrows in the above diagrams, one can define the dual structure of algebra. A $\mathbb{K}$–vector space C together with a co-associative bilinear map $\Delta_C : C \to C \otimes C$ and a counit $\varepsilon : C \longrightarrow \mathbb{K}$ is called *coalgebra*, if we have the following commutative diagrams.

$$\begin{array}{ccc} C \otimes C \otimes C & \xleftarrow{\Delta_C \otimes id_C} & C \otimes C \\ {\scriptstyle id_C \otimes \Delta_C} \uparrow & & \uparrow {\scriptstyle \Delta_C} \\ C \otimes C & \xleftarrow{\Delta_C} & C \end{array} \tag{2.1.4}$$

$$\mathbb{K} \otimes C \xleftarrow{\varepsilon \otimes id_C} C \otimes C \xrightarrow{id_C \otimes \varepsilon} C \otimes \mathbb{K} \tag{2.1.5}$$
with Δ_C from C upward.

This coalgebra is *co-commutative*, if $\tau \circ \Delta = \Delta$. With using the *Sweedler's notation* for the coproduct namely,

$$\Delta x = \sum_{(x)} x_1 \otimes x_2, \tag{2.1.6}$$

the co-associativity and the co-commutativity conditions will be written with

$$(\Delta \otimes id) \circ \Delta(x) = \sum_{(x)} x_{1,1} \otimes x_{1,2} \otimes x_2 = \sum_{(x)} x_1 \otimes x_{2,1} \otimes x_{2,2} = (id \otimes \Delta) \circ \Delta(x), \tag{2.1.7}$$

$$\sum_{(x)} x_1 \otimes x_2 = \sum_{(x)} x_2 \otimes x_1. \tag{2.1.8}$$

The sub-structures such as sub-algebra and sub-coalgebra can be defined in a natural way.

For given algebra A and coalgebra C, one can define a product namely, *convolution product* on the space $L(C, A)$ of all linear maps from C to A. For each φ, ψ in $L(C, A)$, it is given by

$$\varphi * \psi := m_A \circ (\varphi \otimes \psi) \circ \Delta_C. \tag{2.1.9}$$

A $\mathbb{K}$–vector space B together with the unital algebra structure (m, μ) and the counital coalgebra structure (Δ, ε) is called *bialgebra*, if Δ and ε are algebra morphisms and μ is a coalgebra morphism. These conditions are determined with the following commutative diagrams.

$$\begin{array}{ccc} B \otimes B \otimes B \otimes B & \xrightarrow{\tau_{23}} & B \otimes B \otimes B \otimes B \\ {\scriptstyle \Delta_B \otimes \Delta_B} \uparrow & & \downarrow {\scriptstyle m_B \otimes m_B} \\ B \otimes B \xrightarrow{m_B} & B & \xrightarrow{\Delta_B} B \otimes B \end{array} \tag{2.1.10}$$

$$\begin{array}{ccc} B \otimes B \xrightarrow{\varepsilon \otimes \varepsilon} \mathbb{K} \otimes \mathbb{K} & \quad & B \otimes B \xleftarrow{\mu \otimes \mu} \mathbb{K} \otimes \mathbb{K} \\ {\scriptstyle m_B} \downarrow \quad \downarrow {\scriptstyle \sim} & \quad & {\scriptstyle \Delta_B} \uparrow \quad \uparrow {\scriptstyle \sim} \\ B \xrightarrow{\varepsilon} \mathbb{K} & \quad & B \xleftarrow{\mu} \mathbb{K} \end{array} \tag{2.1.11}$$

A bialgebra H together with a linear map $S : H \longrightarrow H$ is called *Hopf algebra*, if there is a compatibility between S and bi-algebraic structure given by the following commutative diagram.

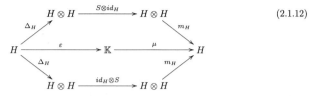 (2.1.12)

The map S is called *antipode*.

There are many examples of Hopf algebras and with attention to our scopes in continue at first we familiar with some important Hopf algebras and reader can find more other samples in [1, 49, 96].

As the first example, for a fixed invertible element $q \in \mathbb{K}$, Hopf algebra H_q is generated by 1 and elements a, b, b^{-1} together with relations

$$bb^{-1} = b^{-1}b = 1, \quad ba = qab.$$

Its structures are determined by

$$\Delta a = a \otimes 1 + b \otimes a, \quad \Delta b = b \otimes b, \quad \Delta b^{-1} = b^{-1} \otimes b^{-1},$$

$$\varepsilon(a) = 0, \quad \varepsilon(b) = \varepsilon(b^{-1}) = 1, \quad S(a) = -b^{-1}a, \quad S(b) = b^{-1}, \quad S(b^{-1}) = b.$$

[79]

For the second example, let G be a finite group and $\mathbb{K}G$ the vector space generated by G. Identify a Hopf algebra structure on $\mathbb{K}G$ such that its product is given by the group structure of G and

$$1 = e, \quad \Delta(g) = g \otimes g,$$

$$\varepsilon(g) = 1, \quad S(g) = g^{-1}. \tag{2.1.13}$$

[79]

For the third example, let $\mathfrak{g}$ be a finite dimensional Lie algebra over $\mathbb{K}$. The universal enveloping algebra $U(\mathfrak{g})$ is the noncommutative algebra generated by 1 and elements of the Lie algebra with respect to the relation

$$[x, y] = xy - yx.$$

It introduces a Hopf algebra structure such that

$$\Delta(x) = x \otimes 1 + 1 \otimes x, \quad \varepsilon(x) = 0, \quad S(x) = -x.$$

One can show that $U(\mathfrak{g})$ is the quotient of the tensor algebra $T(\mathfrak{g})$ modulo an ideal generated by the commutator. [22]

As the last example, consider the algebra of multiple semigroup H of natural positive integers $\mathbb{N}$ such that $(e_n)_{n \in \mathbb{N}}$ is its basis as a vector space. With the help of decomposition of numbers into the prime factors, one can define a commutative cocommutative connected graded (with the number of prime factors (including multiplicities)) Hopf algebra such that its coproduct and its antipode are determined by

$$\Delta(e_{p_1\cdots p_k}) = \sum_{I \sqcup J = \{1,\ldots,k\}} e_{p_I} \otimes e_{p_J},$$

$$S(e_n) = (-1)^{|n|} e_n, \qquad (2.1.14)$$

where p_I denotes the product of the primes $p_j, j \in I$. [78]

By adding some additional structures such as grading and filtration, one can apply Hopf algebras in physics. For example it helps us to classify all Feynman diagrams with respect to loop numbers or number of internal edges and it will be useful when we do renormalization.

Definition 2.1.1. *(i) A Hopf algebra H over $\mathbb{K}$ is called graded, if it is a graded $\mathbb{K}$-vector space $H = \bigoplus_{n \geq 0} H_n$ such that*

$$H_p.H_q \subset H_{p+q}, \quad \Delta(H_n) \subset \bigoplus_{p+q=n} H_p \otimes H_q, \quad S(H_n) \subset H_n.$$

(ii) A connected filtered Hopf algebra H is a $\mathbb{K}$-vector space together with an increasing $\mathbb{Z}_+$-filtration:

$$H^0 \subset H^1 \subset \cdots \subset H^n \subset \cdots, \quad \bigcup_n H^n = H$$

such that H^0 is one dimension and

$$H^p.H^q \subset H^{p+q}, \quad \Delta(H^n) \subset \sum_{p+q=n} H^p \otimes H^q, \quad S(H^n) \subset H^n.$$

[27, 78]

It is easily seen that a graded bialgebra determines an increasing filtration $H^n = \bigoplus_{p=0}^n H_p$. The convolution product together with a filtration structure determine an antipode on a bialgebra. Any connected filtered bialgebra H is a filtered Hopf algebra. Its antipode structure is given by

$$S(x) = \sum_{k \geq 0} (\mu\varepsilon - Id_H)^{*k}(x). \qquad (2.1.15)$$

[78]

For a fixed connected filtered bialgebra H and an algebra A, set a map $e := \mu_A \circ \varepsilon_H$ such that $e(\mathbf{1}) = \mathbf{1}_A$ and $e(x) = 0$ for any $x \in Ker\ \varepsilon_H$. It is easy to see that e plays the role of unit for the convolution product $*$ on the set $L(H, A)$. Set

$$G := \{\varphi \in L(H, A),\ \varphi(\mathbf{1}) = \mathbf{1}_A\}, \qquad (2.1.16)$$

$$\mathfrak{g} := \{\alpha \in L(H, A),\ \alpha(\mathbf{1}) = 0\}. \qquad (2.1.17)$$

It will be shown that elements of these sets make possible to find a new prescription from physical information of a quantum field theory.

Theorem 2.1.1. *(i)* $(G, *)$ *is a group such that for each* $\varphi \in G$, *its inverse is given by*

$$\varphi^{*-1}(x) = \big(e - (e - \varphi)\big)^{*-1}(x) = \sum_{k \geq 0} (e - \varphi)^{*k}(x).$$

(ii) $\mathfrak{g}$ *is a subalgebra of* $(L(H, A), *)$ *such that commutator with respect to the convolution product introduces a Lie algebra structure on it.*
(iii) $G = e + \mathfrak{g}$.
(iv) For any $x \in H^n$, *the exponential map is defined by*

$$e^{*\alpha}(x) = \sum_{k \geq 0} \frac{\alpha^{*k}(x)}{k!}.$$

It determines a bijection map from $\mathfrak{g}$ *onto* G *such that its inverse namely, the logarithmic map is given by*

$$Log(1 + \alpha)(x) = \sum_{k \geq 1} \frac{(-1)^{k-1}}{k} \alpha^{*k}(x).$$

(v) The above sums have just finite terms. [27, 49, 78, 86]

It is also interesting to know that the increasing filtration on H can identify a complete metric structure on $L(H, A)$. Set

$$L^n := \{\alpha \in L(H, A), \alpha|_{H^{n-1}} = 0\}. \tag{2.1.18}$$

It can be seen that for each positive integer numbers p and q,

$$L^p * L^q \subset L^{p+q}. \tag{2.1.19}$$

It gives a decreasing filtration on $L(H, A)$ such that $L^0 = L(H, A)$ and $L^1 = \mathfrak{g}$. For each element $\varphi \in L(H, A)$, the value $val\ \varphi$ is defined as the biggest integer k such that φ is in L^k. The map

$$d(\varphi, \psi) = 2^{-val(\varphi - \psi)} \tag{2.1.20}$$

gives us a complete metric on $L(H, A)$.

We close this section with introducing one important technique for constructing Hopf algebras from Lie algebras namely, Milnor-Moore theorem. Example ?? leads to a closed relation between Lie algebras and Hopf algebras but in general, it is impossible to reconstruct a Hopf algebra from a Lie algebra. By adding some conditions, one can find very interesting process to recover Hopf algebras.

An element p in the Hopf algebra H is called *primitive*, if

$$\Delta(p) = p \otimes 1 + 1 \otimes p. \tag{2.1.21}$$

A graded Hopf algebra is called *finite type*, if each of the homogenous components H_i are finite dimensional vector spaces.

One should mark that if the graded Hopf algebra H (of finite type) is an infinite dimensional vector space, its graded dual

$$H^* = \bigoplus_{n \geq 0} H_n^* \tag{2.1.22}$$

is strictly contained in the space of linear functionals $H^* := L(H, \mathbb{K})$.

Remark 2.1.2. *Let H be a commutative (cocommutative) connected graded finite type Hopf algebra.*
(i) $Ker\ \varepsilon \simeq \bigoplus_{i>0} H_i$ is an ideal in H. It is called augmentation ideal.
(ii) A linear map $f \in H^*$ belongs to H^* if and only if $f|_{H_i} = 0$, for each component H_i but for a finite number.
(iii) If H is finite dimensional vector space, then $H^* = H^\star$.
(iv) There is a cocommutative (commutative) connected graded Hopf algebra structure on H^*.
[27, 49, 78, 86]

For a given Hopf algebra H, an element x in the augmentation ideal is called *indecomposable*, if it can not been written as a linear combination of products of elements in $Ker\ \varepsilon$. The set of all indecomposable elements is denoted by $In(H)$.

Theorem 2.1.2. *For a given connected, graded and finite type Hopf algebra H,*
(i) There is a correspondence between the set of all primitive elements of H namely, $Prim(H)$ and $In(H^*)$.
(ii) There is a correspondence between $In(H)$ and $Prime(H^*)$. [49, 86]

Milnor and Moore proved that the reconstruction of a given Hopf algebra (under some conditions) on the basis of primitive elements is possible. In fact, this result plays an essential role in the Connes-Marcolli categorical configuration in the study of renormalizable quantum field theories.

Theorem 2.1.3. *Let H be a connected graded commutative finite type Hopf algebra. It can be reconstructed with the Lie subalgebra $Prim(H)$ of $L(H, \mathbb{K})$ and it means that $H \simeq U(Prim(H))^*$. [22, 86]*

Two classes of elements in H^* have particular roles namely, *characters* and *infinitesimal characters*. It is discussed by Kreimer that Feynman rules of a given quantum field theory can be capsulated in characters where this ability provides a new reformulation from counterterms, renormalized values, elements of the renormalization group and its related infinitesimal generator (β–function).

An element $f \in H^*$ is called character, if $f(1) = 1$ and for each $x, y \in H$,

$$f(xy) = f(x)f(y). \tag{2.1.23}$$

An element $g \in H^*$ is called derivation (or infinitesimal character), if for each $x, y \in H$,

$$g(xy) = g(x)\varepsilon(y) + g(y)\varepsilon(x). \tag{2.1.24}$$

It is important to note that each primitive element of the Hopf algebra H^* determines a derivation.

2.2 Combinatorial Hopf algebras

Hopf algebra of renormalization is introduced on the set of all Feynman diagrams of a given renormalizable physical theory such that its structures completely related to the renormalization process on these graphs. Since we want to have a general framework to consider perturbation theory, so it is necessary to identify a Hopf algebra structure independent of physical theories and further, it is reasonable to provide a universal simplified toy model for this Hopf algebraic formalization to apply it in computations. Fortunately, investigation of the combinatorics of the renormalization can help us to find a solution for this problem. Indeed, Kreimer applied decorated version of non-planar rooted trees (as combinatorial objects) to represent Feynman diagrams such that labels could help us to restore divergent sub-diagrams and their positions (i.e. nested loops) in origin graphs. Then with respect to the recursive mechanism for removing sub-divergences, he introduced a coproduct structure on these labeled rooted trees such that as the result one can produce a combinatorial Hopf algebra independent of physical theories. It is called Connes-Kreimer Hopf algebra of rooted trees [52, 58]. Even more in a categorical configuration, this rooted tree type model is equipped with a universal property with respect to Hochschild cohomology theory such that the grafting operator can determine its related Hochschild one cocycles [28, 29, 32]. At this level one can expect the applications of combinatorial techniques in the study of perturbative renormalization [17, 31, 37, 43, 44, 45, 46, 47, 50, 71].

In this chapter we focus on combinatorial objects and review the structures of some important defined combinatorial Hopf algebras which are connected with the Connes-Kreimer Hopf algeba.

2.2.1 Connes-Kreimer Hopf algebra

Rooted trees allow us to investigate the combinatorial basement of renormalization programme such that as one expected result, it determines a universal simplification for explaining the removing of sub-divergences procedure. Here we consider the most important Hopf algebra structure on non-planar rooted trees in the study of QFTs.

Definition 2.2.1. *A non-planar rooted tree t is an oriented, connected and simply connected graph together with one distinguished vertex with no incoming edge namely, root. A monomial in rooted trees (that commuting with each other) is called forest.*

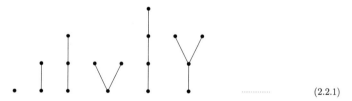

(2.2.1)

A rooted tree t with a given embedding in the plane is called planar rooted tree. For example,

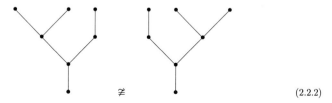

(2.2.2)

17

Let **T** be the set of all non-planar rooted trees and $\mathbb{K}\mathbf{T}$ be the vector space over the field $\mathbb{K}$ (with characteristic zero) generated by **T**. It is graded by the number of non-root vertices of rooted trees and it means that

$$\mathbf{T_n} := \{t \in \mathbf{T} : |t| = n+1\}, \quad \mathbb{K}\mathbf{T} := \bigoplus_{n \geq 0} \mathbb{K}\mathbf{T_n}. \tag{2.2.3}$$

Consider graded free unital commutative symmetric algebra $H(\mathbf{T})$ containing $\mathbb{K}\mathbf{T}$ such that the empty tree is its unit. We equip this space with the counit $\epsilon : H(\mathbf{T}) \longrightarrow \mathbb{K}$ given by

$$\epsilon(\mathbb{I}) = 1, \quad \epsilon(t_1...t_n) = 0, \quad t_1...t_n \neq \mathbb{I}. \tag{2.2.4}$$

With respect to the BPHZ renormalization process, some edges and vertices from rooted trees should be removed step by step and this can be formulated with a special family of cuts.

Definition 2.2.2. *An admissible cut c of a rooted tree t is a collection of its edges with this condition that along any path from the root to the other vertices, it meets at most one element of c. By removing the elements of an admissible cut c from a rooted tree t, we will have a rooted tree $R_c(t)$ with the original root and a forest $P_c(t)$ of rooted trees.*

For instance,

(2.2.5)

shows an admissible cut but the cut

(2.2.6)

can not be admissible. This concept determines a coproduct structure on $H(\mathbf{T})$ given by

$$\Delta : H(\mathbf{T}) \longrightarrow H(\mathbf{T}) \otimes H(\mathbf{T}), \quad \Delta(t) = t \otimes \mathbb{I} + \mathbb{I} \otimes t + \sum_c P_c(t) \otimes R_c(t) \tag{2.2.7}$$

where the sum is over all possible non-trivial admissible cuts on t. As an example,

(2.2.8)

It should be remarked that this coproduct can be rewritten in a recursive way. Let B^+ : $H(\mathbf{T}) \longrightarrow H(\mathbf{T})$ be a linear operator that mapping a forest to a rooted tree by connecting the roots of rooted trees in the forest to a new root.

$$\text{(2.2.9)}$$

Operator B^+ is an isomorphism of graded vector spaces and for the rooted tree $t = B^+(t_1...t_n)$, we have

$$\Delta B^+(t_1...t_n) = t \otimes \mathbb{I} + (id \otimes B^+)\Delta(t_1...t_n). \tag{2.2.10}$$

Δ is extended linearity to define it as an algebra homomorphism. On the other hand, with the help of admissible cuts one can define recursively an antipode on $H(\mathbf{T})$ given by

$$S(t) = -t - \sum_c S(P_c(t))R_c(t). \tag{2.2.11}$$

Theorem 2.2.1. *The symmetric algebra $H(\mathbf{T})$ together with the coproduct (2.2.7) and the antipode (2.2.11) is a finite type connected graded commutative noncocommutative Hopf algebra. It is called Connes-Kreimer Hopf algebra and denoted by H_{CK}. [13, 27, 28, 29]*

The study of Hopf subalgebras of H_{CK} can be useful. For instance one can consider the cocommutative Hopf subalgebra of *ladder trees* (i.e. rooted trees without any side-branchings) $H(\mathbf{LT})$ such that it is applied to work on the relations between perturbative QFTs and representation theory of Lie algebras. For this case, $H(\mathbf{LT})$ is reduced to a polynomial algebra freely generated by ladder trees and with the help of increasing or decreasing the degree of generators, one can induce insertion and elimination operators. [84, 85]

By theorem 2.1.1, the convolution product $*$ determines a group structure on the space $\mathbf{char H_{CK}}$ of all characters and a graded Lie algebra structure on the space $\partial\mathbf{char H_{CK}}$ of all derivations where naturally, there is a bijection map $\exp^*$ from $\partial\mathbf{char H_{CK}}$ to $\mathbf{char H_{CK}}$ (which plays an essential role in the representation of components of the Birkhoff decomposition of characters). [28, 29]

Finally one should mark to the *universal property* of this Hopf algebra such that it is the essential result of the universal problem in Hochschild cohomology.

Theorem 2.2.2. *Let $\mathcal{C}$ be a category with objects (H, L) consisting of a commutative Hopf algebra H and a Hochschild one cocycle $L : H \longrightarrow H$. It means that for each $x \in H$,*

$$\Delta L(x) = L(x) \otimes \mathbb{I} + (id \otimes L)\Delta(x).$$

And also Hopf algebra homomorphisms, that commute with cocycles, are morphisms in this category. (H_{CK}, B^+) is the universal element in $\mathcal{C}$. In other words, for each object (H, L) there exists a unique morphism of Hopf algebras $\phi : H_{CK} \longrightarrow H$ such that $L \circ \phi = \phi \circ B^+$. H_{CK} is unique up to isomorphism. [13]

2.2.2 Rooted trees and (quasi-)symmetric functions

There are different Hopf algebra structures on (non-)planar rooted trees and in fact, Connes-Kreimer Hopf algebra is one particular choice. Here we try to familiar with some important combinatorial Hopf algebras and then with using (quasi-)symmetric functions, their relations with H_{CK} will be considered.

Definition 2.2.3. *Let t, s be rooted trees such that $t = B^+(t_1, ...t_n)$ and $|s| = m$. The new product $t \bigcirc s$ is defined with the sum of rooted trees given by attaching each of t_i to a vertex of s.*

One can define a coproduct compatible with $\bigcirc$ on $\mathbb{K}\mathbf{T}$. It is given by

$$\Delta_{GL} B^+(t_1,...t_k) = \sum_{I \cup J = \{1,2,...,k\}} B^+(t(I)) \otimes B^+(t(J)). \qquad (2.2.12)$$

Theorem 2.2.3. $H_{GL} := (\mathbb{K}\mathbf{T}, \bigcirc, \Delta_{GL})$ *is a connected graded noncommutative cocommutative Hopf algebra and it is called Grossman-Larson Hopf algebra. H_{GL} is the graded dual of H_{CK} and it is the universal enveloping algebra of its Lie algebra of primitives. [43, 87]*

Let $\mathbf{P}$ be the set of all planar rooted trees and $\mathbb{K}\mathbf{P}$ be its graded vector space. Tensor algebra $T(\mathbb{K}\mathbf{P})$ is an algebra of ordered forests of planar rooted trees and $B^+ : T(\mathbb{K}\mathbf{P}) \longrightarrow \mathbb{K}\mathbf{P}$ is an isomorphism of graded vector spaces. There are two interesting Hopf algebra structures on $\mathbf{P}$.

Definition 2.2.4. *A balanced bracket representation (BBR) of a planar rooted tree contains symbols $<$ and $>$ satisfying in the following rules:*
- *For a planar rooted tree of weight n, the symbol $<$ and the symbol $>$ occur n times,*
- *In reading from left to right, the count of $<$'s is agree with the count of $>$'s,*
- *The empty BBR is a tree with just one vertex.*

For example, one represents planar rooted trees

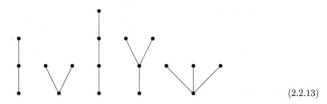

(2.2.13)

with

$$<<>>, \quad <><>, \quad <<<>>>, \quad <<><>>, \quad <><><>, \qquad (2.2.14)$$

respectively.

Definition 2.2.5. *A BBR F is called irreducible, if $F =< G >$ for some BBR G and otherwise it can be written by a juxtaposition $F_1...F_k$ of irreducible BBRs. These components correspond with the branches of the root in the associated planar rooted tree.*

Definition 2.2.6. *Let t, s be two planar rooted trees with BBR representations F_t, F_s such that $F_t = F_t^1...F_t^k$. Define a new product $t \diamond s$ by a sum of planar rooted trees such that their BBRs are given by shuffling the components of F_t into the F_s.*

Moreover, with help of the decomposition of elements into their irreducible components, one can modify a compatible coproduct $\Delta_\diamond$ on $\mathbb{K}\mathbf{P}$.

Theorem 2.2.4. *(i) Based on the balanced bracket representation, there is a connected graded noncommutative Hopf algebra structure on $\mathbb{K}\mathbf{P}$ and it is denoted by $H_\mathbf{P} := (\mathbb{K}\mathbf{P}, \diamond, \Delta_\diamond)$.*
(ii) Based on the coproduct (2.2.7), there is a graded connected noncommutative Hopf algebra. It is called Foissy Hopf algebra and denoted by H_F. H_F is self-dual and isomorphic to $H_\mathbf{P}$. [35, 44, 47]

Relation between rooted trees and noncommutative geometry can be clear when the reconstruction of one important Hopf algebra in computations of transverse index theory, based on rooted trees, is done [13, 50]. Consider a Hopf algebra H with the generators x, y, δ_n $(n \in \mathbb{N})$, together with the following relations

$$[x,y] = -x, \quad [x, \delta_n] = \delta_{n+1}, \quad [y, \delta_n] = n\delta_n, \quad [\delta_n, \delta_m] = 0, \tag{2.2.15}$$

such that its coproduct structure on the generators is given by

$$\Delta(x) = x \otimes 1 + 1 \otimes x + \delta_1,$$
$$\Delta(y) = y \otimes 1 + 1 \otimes y,$$
$$\Delta(\delta_1) = \delta_1 \otimes 1 + 1 \otimes \delta_1. \tag{2.2.16}$$

It is easy to present the generators δ_n with rooted trees. Define a linear operator N on rooted trees such that its application on a rooted tree t is a sum of rooted trees given by adding an edge to each vertex of t. Now identify δ_1 with $\bullet$ and $\delta_n = N^n(\mathbb{I})$.

Lemma 2.2.7. *The set of generators $\{\delta_n\}_n$ introduces a Hopf subalgebra of H such that it is equivalent to the Connes-Moscovici Hopf algebra H_{CM}. [13]*

It will be shown that how one can reduce H_{CM} based on Dyson-Schwinger equations and additionally, a copy of this Hopf algebra related to the universal Hopf algebra of renormalization will be determined. These results report the importance of this Hopf algebra in the study of quantum field theory.

Definition 2.2.8. *Let $\mathbb{K}[[x_1, x_2, ...]]$ be the ring of formal power series. A formal series f is called*
(i) symmetric, if for any sequence of distinct positive integers $n_1, ..., n_k$, the coefficients in f of the monomials $x_{n_1}^{i_1}...x_{n_k}^{i_k}$ and $x_1^{i_1}...x_k^{i_k}$ equal.
(ii) quasi-symmetric, if for any increasing sequence $n_1 < ... < n_k$, the coefficients in f of the monomials $x_{n_1}^{i_1}...x_{n_k}^{i_k}$ and $x_1^{i_1}...x_k^{i_k}$ equal.
(iii) Let SYM $(QSYM)$ be the set of all symmetric (quasi-symmetric) functions. It is easy to see that $SYM \subset QSYM$.

For better understanding, it can be seen that for each n the symmetric group $\mathbb{S}_n$ acts on $\mathbb{K}[[x_1, x_2, ...]]$ by permuting the variables and a symmetric function is invariant under these actions and it means that after each permutation coefficients of its monomials remain without any change.

Lemma 2.2.9. *(i) As a vector space, $QSYM$ is generated by the monomial quasi-symmetric functions M_I such that $I = (i_1, ..., i_k)$ and $M_I := \sum_{n_1 < n_2 < ... < n_k} x_{n_1}^{i_1}...x_{n_k}^{i_k}$.*
(ii) If we forget order in a composition, then the generators $m_\lambda := \sum_{\phi(I) = \lambda} M_I$ of SYM (viewed as a vector space) can be determined. [41, 44]

Theorem 2.2.5. *(i) There is a graded connected commutative cocommutative self-dual Hopf algebra structure on SYM.*
(ii) There is a graded connected commutative non-cocommutative Hopf algebra structure on $QSYM$ such that its graded dual is denoted by $NSYM$. As an algebra, $NSYM$ is the noncommutative polynomials on the variables z_n of degree n. [41, 44]

Hoffman could find new important relations between rooted trees and (quasi-)symmetric functions such that we will extend his results to the level of the universal Hopf algebra of renormalization.

Theorem 2.2.6. *There are following commutative diagrams of Hopf algebra homomorphisms. [44]*

$$\begin{array}{ccccccc} NSYM & \xrightarrow{\alpha_1} & H_F & & SYM & \xleftarrow{\alpha_4^\star} & H_{GL} \\ {\scriptstyle \alpha_3}\downarrow & & {\scriptstyle \alpha_2}\downarrow & & {\scriptstyle \alpha_3^\star}\downarrow & & {\scriptstyle \alpha_2^\star}\downarrow \\ SYM & \xrightarrow{\alpha_4} & H_{CK} & & QSYM & \xleftarrow{\alpha_1^\star} & H_P \end{array} \qquad (2.2.17)$$

Proof. It is enough to define homomorphisms on generators. With attention to the definitions of Hopf algebras, we have
- α_1 sends each variable z_n to the ladder tree l_n of degree n.
- α_2 maps each planar rooted tree to its corresponding rooted tree without notice to the order in products.
- α_3 sends each z_n to the symmetric function $m_{\underbrace{(1,...,1)}_{n}}$.
- α_4 maps $m_{\underbrace{(1,...,1)}_{n}}$ to the ladder tree l_n.
- For the composition $I = (i_1, ..., i_k)$ define a planar rooted tree $t_I := B^+(l_{i_1}, ..., l_{i_k})$. For each planar rooted tree t, if $t = t_I$, then define $\alpha_1^\star(t) := M_I$ and otherwise $\alpha_1^\star(t) := 0$.
- For each rooted tree t, $\alpha_2^\star(t) := |sym(t)| \sum_{s \in \alpha_2^{-1}(t)} s$.
- $\alpha_3^\star$ is the inclusion map.
- For the partition $J = (j_1, ..., j_k)$, define a rooted tree $t_J := B^+(l_{j_1}, ..., l_{j_k})$. For each rooted tree t, if $t = t_J$ (for some partition J), then define $\alpha_4^\star(t) := |sym(t_J)| m_J$ and otherwise $\alpha_4^\star(t) := 0$. □

Definition 2.2.10. *Recursively define the following morphism*

$$Z: NSYM \longrightarrow H_{GL}, \quad Z(z_n) = \epsilon_n$$

such that rooted trees ϵ_n are given by

$$\epsilon_0 := \bullet$$

$$\epsilon_n := k_1 \bigcirc \epsilon_{n-1} - k_2 \bigcirc \epsilon_{n-2} + ... + (-1)^{n-1} k_n$$

where

$$k_n := \sum_{|t|=n+1} \frac{t}{|sym(t)|} \in H_{GL}.$$

It is called Zhao's homomorphism.

From definition it is clear that Z is an injective homomorphism of Hopf algebras.

Lemma 2.2.11. *Dual of Zhao's homomorphism exists uniquely. [45, 46]*

Proof. Suppose
$$A^+ : QSYM \longrightarrow QSYM, \quad M_I \longmapsto M_{I \sqcup (1)}.$$
It is a linear map with the cocycle property. For each ladder tree l_n and monomial u of rooted trees, define a morphism $Z^\star : H_{CK} \longrightarrow QSYM$ such that
$$l_n \longmapsto M_{\underbrace{(1,...,1)}_{n}},$$
$$B^+(u) \longmapsto A^+(Z^\star(u)).$$
One can see that $Z^\star$ is the unique homomorphism with respect to the map A^+. □

We show that it is possible to lift the Zhao's homomorphism and its dual to the level of Hall rooted trees and Lyndon words. Roughly, this process provides an extension of this homomorphism to the level of the universal Hopf algebra of renormalization.

2.2.3 Incidence Hopf algebras

On the one hand, rooted trees introduce one important class of Hopf algebras namely, combinatorial type and on the other hand, incidence Hopf algebras, induced in operad theory, provide another general class of Hopf algebras such that rooted trees (as kind of posets) characterize interesting examples in this procedure. The essential part of this story is that incidence Hopf algebra related to one special family of operads introduces the Connes-Kreimer Hopf algebra.

The story of operad theory was begun with the study of loop spaces and then its applications in different branches of mathematics were found very soon. There is a closed relation between operads and objects of symmetric monoidal categories such as category of sets, category of topological spaces, category of vector spaces, ... Additionally, operads can determine interesting source of Hopf algebras namely, incidence Hopf algebras [17, 92, 100]. In this part we are going to consider this important family of Hopf algebras related to posets.

Definition 2.2.12. *A partially ordered set (poset) is a set with a partial order relation. A growing sequence of the elements of a poset is called chain. A poset is pure, if for any $x \leq y$ the maximal chains between x and y have the same length. A bounded and pure poset is called graded poset.*

Example 2.2.13. *One can define a graded partial order on the set $[n] = \{1, 2, ..., n\}$ by the refinement of partitions and it is called partition poset.*

Definition 2.2.14. *(i) An operad $(P, \mathbf{co}, \mathbf{u})$ is a monoid in the monoidal category $\mathbb{S} - Mod$ of $\mathbb{S}-modules$ (i.e. a collection $\{P(n)\}_n$ of (right) $\mathbb{S}_n-modules$). It means that the composition morphism $\mathbf{co} : P \circ P \longrightarrow P$ is associative and the morphism $\mathbf{u} : \mathbb{I} \longrightarrow P$ is unit.*
(ii) This operad is called augmented, if there exists a morphism of operads $\psi_P : P \longrightarrow \mathbb{I}$ such that $\psi_P \circ \mathbf{u} = id$.

Example 2.2.15. *A $\mathbb{S}-set$ is a collection $\{P_n\}_n$ of sets P_n equipped with an action of the group $\mathbb{S}_n$. A monoid $(P, \mathbf{co}, \mathbf{u})$ in the monoidal category of $\mathbb{S}-sets$ is called a set operad.*

Definition 2.2.16. *For a given set operad P and for each $(x_1, ..., x_t) \in P_{i_1} \times ... \times P_{i_t}$, define a map*
$$\lambda_{(x_1,...,x_t)} : P_t \longrightarrow P_{i_1+...+i_t}, \quad x \longmapsto \mathbf{co}(x \circ (x_1, ..., x_t)).$$
A set opeard P is called basic, if each $\lambda_{(x_1,...,x_t)}$ be injective.

Incidence Hopf algebras are introduced based on special partition posets associated to set operads. At first we should know the structure of this class of posets.

Definition 2.2.17. *Suppose $(P, \mathbf{co}, \mathbf{u})$ be a set operad and for the given set A with n elements, let $\mathbb{A}$ be the set of ordered sequences of the elements of A such that each element appearing once. For each n, there is an action of the group $\mathbb{S}_n$ on P_n such that for each element $x_n \times (a_{i_1}, ..., a_{i_n})$ in $P_n \times \mathbb{A}$, its image under an element σ of $\mathbb{S}_n$ is given by*
$$\sigma(x_n) \times (a_{\sigma^{-1}(i_1)}, ..., a_{\sigma^{-1}(i_n)}).$$
It is called diagonal action and its orbit is denoted by $\overline{x_n \times (a_{i_1}, ..., a_{i_n})}$.

Definition 2.2.18. *Let $\mathfrak{P}_n(A) := P_n \times_{\mathbb{S}_n} \mathbb{A}$ be the set of all orbits under the diagonal action. Set*
$$P(A) := (\bigsqcup_{f:[n] \longrightarrow A} P_n)_\sim$$
where f is a bijection and $(x_n, f) \sim (\sigma(x_n), f \circ \sigma^{-1})$ is an equivalence relation. A $P-$partition of $[n]$ is a set of components $B_1, ..., B_t$ such that
- *Each B_j belongs to $\mathfrak{P}_{i_j}(I_j)$ where $i_1 + ... + i_t = n$,*
- *Family $\{I_j\}_{1 \leq j \leq t}$ is a partition of $[n]$.*

Lemma 2.2.19. *One can extend maps $\lambda_{(x_1,...,x_t)}$ to $\lambda^\sim$ at the level of $P(A)$. [100]*

Proof. Define
$$\lambda^\sim : P_t \times (\mathfrak{P}_{i_1}(I_1) \times ... \times \mathfrak{P}_{i_t}(I_t)) \longrightarrow \mathfrak{P}_{i_1+...+i_t}(A)$$
$$x \times (c_1, ..., c_t) \longmapsto \overline{\mathbf{co}(x \circ (x_1, ..., x_t))} \times (a_1^1, ..., a_{i_t}^t)$$
such that $\{I_j\}_{1 \leq j \leq t}$ is a partition of A and each c_r is represented by $\overline{x_r \times (a_1^r, ..., a_{i_r}^r)}$ where $x_r \in P_{i_r}$, $I_r = \{a_1^r, ..., a_{i_r}^r\}$. □

Definition 2.2.20. *For the set operad P and $P-$partitions $\mathfrak{B} = \{B_1, ..., B_r\}$, $\mathfrak{C} = \{C_1, ..., C_s\}$ of $[n]$ such that $B_k \in \mathfrak{P}_{i_k}(I_k)$ and $C_l \in \mathfrak{P}_{j_l}(J_l)$, we say that the $P-$partition $\mathfrak{C}$ is larger than $\mathfrak{B}$, if for any $k \in \{1, 2, ..., r\}$ there exists $\{p_1, ..., p_t\} \subset \{1, 2, ..., s\}$ such that*
- *Family $\{J_{p_1}, ..., J_{p_t}\}$ is a partition of I_k,*
- *There exists an element $x_t \in P_t$ such that $B_k = \lambda^\sim(x_t \times (C_{p_1}, ..., C_{p_t}))$.*

This poset is called operadic partition poset associated to the operad P and denoted by $\Pi_P([n])$.

One can develop the notion of this poset to each locally finite set $A = \bigsqcup A_n$ such that in this case a $P-$partition of $[A]$ is a disjoint union (composition) of $P-$partitions of $[A_n]$s and therefore the operadic partition poset associated to the operad P will be a composition of posets $\Pi_P([A_n])$ and denoted by $\Pi_P([A])$.

Definition 2.2.21. *A collection* $(\mathfrak{p}_i)_{i \in I}$ *of posets is called good collection, if*
- *Each poset $\mathfrak{p}_i$ has a minimal element $\mathbf{0}$ and a maximal element $\mathbf{1}$ (an interval),*
- *For all $i \in I$, $x \in \mathfrak{p}_i$, the interval $[\mathbf{0}, x]$ (or $[x, \mathbf{1}]$) is isomorphic to a product of posets $\prod_j \mathfrak{p}_j$ (or $\prod_k \mathfrak{p}_k$).*

Remark 2.2.22. *For a given good collection $\mathcal{A} := (\mathfrak{p}_i)_{i \in I}$, it is possible to make a new good collection $\mathcal{A}^-$ of all finite products $\prod_i \mathfrak{p}_i$ of elements such that it is closed under products and closed under taking subintervals. [17]*

Let $[\mathcal{A}]$ ($[\mathcal{A}^-]$) be the set of isomorphism classes of posets in $\mathcal{A}$ ($\mathcal{A}^-$) such that elements in these sets denoted by $[i], [j], ...$ and $H_{\mathcal{A}}$ be a vector space generated by the family $\{F_{[i]}\}_{[i] \in [\mathcal{A}^-]}$. It is equipped with a commutative product (i.e. direct product of posets) $F_{[i]} F_{[j]} = F_{[i \times j]}$ such that $F_{[e]}$ is the unit (where $[e]$ is the isomorphism class of the singleton interval).

Remark 2.2.23. *As an algebra $H_{\mathcal{A}}$ may not be free.*

Lemma 2.2.24. *Based on subintervals, there is a coproduct structure on $H_{\mathcal{A}}$ given by*

$$\Delta(F_{[i]}) = \sum_{x \in \mathfrak{p}_i} F_{[\mathbf{0}, x]} \otimes F_{[x, \mathbf{1}]}.$$

It determines a commutative Hopf algebra.

Theorem 2.2.7. *Let Π_P be a family of the operadic partition posets associated to the set operad P. One can find a good collection of posets $(\mathfrak{p}_i)$ (depended upon Π_P) such that its related Hopf algebra H_P is called incidence Hopf algebra. [17, 92]*

Remark 2.2.25. *Incidence Hopf algebra H_P has a basis indexed by the isomorphism classes of intervals in the posets $\Pi_P(I)$ (for all sets I) and this identification makes the sets I disappear and it means that the construction of this Hopf algebra is independent of any label.*

A rooted tree looks like a poset with a unique minimal element (root) such that for any element v, the set of elements descending v forms a chain (i.e. the graph has no loop) and maximal elements are called *leaves*. There is an interesting basic set operad on rooted trees such that its incidence Hopf algebra determines a well known object.

Definition 2.2.26. *For the set I with the partition $\{J_i\}_{i \geq 1}$, suppose $NAP(I)$ be the set of rooted trees with vertices labeled by I. For $s_i \in NAP(J_i)$ and $t \in NAP(I)$, we consider the disjoint union of the rooted trees s_i such that for each edge of t between i_1, i_2 in I, add an edge between the root of s_{i_1} and the root of s_{i_2}. The resulting graph is a rooted tree labeled by $\bigsqcup_i J_i$ and its root is the root of s_k such that k is the label of the root of t. It defines the composition $t((s_i)_{i \in I})$.*

Theorem 2.2.8. *Operad NAP is a functor from the groupoid of sets to the category of sets. [17, 92, 100]*

The operadic partition poset $\Pi_{NAP}(I)$ is a set of forests of I-labeled rooted trees such that a forest X is covered by a forest Y, if Y is obtained from X by grafting the root of one component of X to the root of another component of X. Or X is obtained from Y by removing an edge incident to the root of one component of Y.

Remark 2.2.27. *Any interval in $\Pi_{NAP}(I)$ is a product of intervals of the form $[0, t_i]$ such that $t_i \in NAP(J_i)$. If $t = B^+(t_1, ..., t_k)$, then the poset $[0, t]$ is isomorphic to the product of the posets $[0, B^+(t_i)]$ for $i \in \{1, 2, ..., k\}$.*

Lemma 2.2.28. *The incidence Hopf algebra H_{NAP} is a free commutative algebra on unlabeled rooted trees of root-valence 1 such that elements $F_{[t]}$ (where t is a rooted tree) form a basis at the vector space level.*

According to the theorem 2.2.2 and the structure of H_{NAP}, one can obtain the next important result.

Theorem 2.2.9. *H_{NAP} is isomorphic to H_{CK} by the unique Hopf algebra isomorphism ρ : $F_{[B^+(t_1,...,t_k)]} \longmapsto t_1...t_k$. [17]*

This theorem allows us to discover an operadic partition poset formalism for the Connes-Kreimer Hopf algebra of rooted trees such that after finding a rooted tree reformulation for H_U, one can apply theorem 2.2.9 to recognize an operadic source for this specific universal Hopf algebra in Connes-Marcolli treatment.

Chapter 3

Perturbative Renormalization

The initial motivation in collaboration between the theory of Hopf algebras and the perturbation theory in renormalizable QFT was determined carefully with the description of perturbative renormalization underlying dimensional regularization in minimal subtraction scheme in an algebro-geometric framework. In other words, Connes and Kreimer discovered an interesting revolutionary bridge between the BPHZ prescription in renormalization and the Riemann-Hilbert correspondence. They proved that perturbative renormalization is in fact one special case of the general mathematical process of the extraction of finite values based on the Riemann-Hilbert problem in the reconstruction of differential equations from data of their monodromy representation such that for the algebraic reformalization of the BPHZ method, one can look at to the local regular-singular version of this problem where at this level the application of the Birkhoff factorization in the study of QFTs can be investigated. Because in fact, negative part of this decomposition can be applied to correct the behavior of solutions near singularities without introducing new singularities. [14, 15, 20, 21, 22, 40]

According to this mathematical mechanism, for a given renormalizable QFT Φ one can associate an infinite dimensional complex Lie group $G(\mathbb{C})$ (i.e. Lie group of diffeographisms of the theory) determined with the Hopf algebra H_{FG} of Feynman diagrams of the theory and depended on the chosen regularization method (i.e. a commutative algebra A). It should be noticed that since Hopf algebra H_{FG} is graded and finite type therefore the group $G(\mathbb{C})$ is pro-unipotent.

With working on the dimensional regularization in the minimal subtraction scheme, one can find an algebro-geometric machinery to consider perturbative renormalization. It means that each character carries a geometric meaning in the sense that instead of working on characters one can reproduce physical information of a given theory Φ from factorization of loops (which are depended on the mass parameter μ and the dimensional regularization parameter z) with values in the Lie group $G(\mathbb{C})$.

For instance in [14, 15], authors show that passing from unrenormalized value to the renormalized value is equivalent to the replacement of a given loop $z \longmapsto \gamma_\mu(z) \in G(\mathbb{C})$ on the infinitesimal punctured disk Δ^* (identified by the regularization parameter) with the value of its positive component of the Birkhoff decomposition at the critical integral dimension D. In addition, one can recover the related counterterm from the negative part of this decomposition.

These results strongly depend on this essential fact that each regularized unrenormalized value

$U_\mu^z(\Gamma(p_1, ..., p_n))$ determines a loop $\gamma_\mu(z)$ on $\mathbf{\Delta}^*$ around the origin and with values in $G(\mathbb{C})$ such that with the minimal subtraction this unrenormalized value for different values z will be subtracted. [19, 20]

In this chapter, with a pedagogical intention, we are going to consider the Connes-Kreimer approach to renormalization to provide enough knowledge about this new Hopf algebraic interpretation from physical information. We start with the definition of the Hopf algebra of Feynman diagrams and then consider some of its properties such as grading structures, gluing operator, its rooted tree type representation. Finally, Hopf algebraic renormalization will be studied.

3.1 Insertion operator: From a pre-Lie algebra structure on Feynman diagrams to a Hopf algebra

A renormalizable perturbative quantum field theory can be introduced based on a family of graphs namely, Feynman diagrams which describe possible circumstances between different types of elementary particles. In these graphs vertices report interactions and edges indicate propagators. Here one can see some examples of different types of vertices and edges in 3-dimensional scalar field theory, QCD, QED and Gravity:

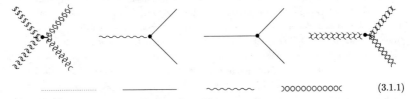

(3.1.1)

Consider a theory with the set R_V (consists of all possible interactions) and the set R_E (consists of all propagators).

Definition 3.1.1. *A Feynman diagram Γ is an oriented graph that contains a finite set Γ^0 of vertices and a finite set Γ^1 of edges such that*
- For each vertex v, its type is determined by the set

$$f_v := \{e \in \Gamma^1 : e \cap v \neq \emptyset\}.$$

- The set Γ^1 decomposes into two different subsets
(Int) Γ_{int}^1 consists of all internal edges (i.e. an edge together with begin and end vertices),
(Ext) Γ_{ext}^1 consists of all external edges (i.e. an edge with an open end).
- Based on Feynman rules of a theory, all edges are labeled with physical parameters (i.e. momenta of particles).
- If $p_1, ..., p_k$ are momenta of external edges, then $\sum_i p_i = 0$. (conservation law)

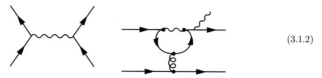
(3.1.2)

A special class of these diagrams together with an algebraic operation (i.e. insertion) are enough to construct the whole theory. They are one particle irreducible (1PI) Feynman graphs without any sub-divergences which play the role of building blocks for defining a mathematical structure (i.e. Hopf algebra).

Definition 3.1.2. *An n-particle irreducible (n-PI) graph is a Feynman diagram Γ with this property that upon removal of n internal edges, it is still connected. It is clear that for $n \geq 2$, each n-PI graph is a (n-1)-PI.*

(3.1.3)

1PI   not 1PI

Definition 3.1.3. *For each arbitrary Feynman diagram Γ,*
(i) $res(\Gamma)$ is a new graph as the result of shrinking all of the internal edges and vertices of Γ into one vertex. The resulting graph consists of a vertex together with all of the external edges of Γ.
(ii) For each Feynman subgraph γ of Γ, the graph Γ/γ is defined by shrinking γ into a vertex. The resulting diagram is called quotient graph.

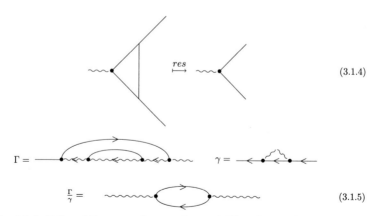

(3.1.4)

(3.1.5)

Definition 3.1.4. *Define a bilinear operation $\star$ on the set of 1PI graphs given by*

$$\Gamma_1 \star \Gamma_2 := \sum_\Gamma n(\Gamma_1, \Gamma_2; \Gamma) \Gamma$$

29

where the sum is over 1PI graphs Γ and $n(\Gamma_1, \Gamma_2; \Gamma)$ counts the number of ways that a subgraph Γ_2 can be reduced to a point in Γ such that Γ_1 is obtained and also

$$|\Gamma| = |\Gamma_1| + |\Gamma_2|, \ res(\Gamma) = res(\Gamma_1).$$

(3.1.6)

(3.1.7)

Remark 3.1.5. *(i) Finitely of Feynman diagrams show that the above sum is finite.*
(ii) $res(\Gamma_1 \star \Gamma_2) = res(\Gamma_1)$,
(iii) The operation $\star$ is pre-Lie, namely

$$[\Gamma_1 \star \Gamma_2] \star \Gamma_3 - \Gamma_1 \star [\Gamma_2 \star \Gamma_3] = [\Gamma_1 \star \Gamma_3] \star \Gamma_2 - \Gamma_1 \star [\Gamma_3 \star \Gamma_2],$$

(iv) For some integers r and k_j that $j = 1, ..., r$, any non-primitive 1PI graph Γ can be written at most in r different forms

$$\Gamma = \prod_{i=1}^{k_j} \gamma_j \star_{j,i} \Gamma_{j,i}$$

such that $\gamma_j s$ are primitive graphs. When $r > 1$, the graph Γ is called overlapping. [57, 62, 63]

So this operator determines a Lie algebra structure on Feynman diagrams such that the Lie bracket is the commutator with respect to the $\star$. From physical point of view, this insertion operator technically can be expounded by the gluing of Feynman graphs based on types of edges. The readers who are interested in this quest of a deeper level of understanding should consult [13, 16, 57].

Theorem 3.1.1. *Graded dual of the universal enveloping algebra of the Lie algebra $\mathcal{L}$ on 1PI graphs (determined with the definition 3.1.4) is a graded connected commutative non-cocommutative Hopf algebra. It is called Hopf algebra of Feynman diagrams of the theory Φ and denoted by $H_{FG} = H(\Phi)$. [19, 22, 52, 58]*

Proof. It is the immediate result of the Milnor-Moore theorem. Based on the gluing information, one can determine sub-diagrams of Feynman graphs such that it leads to the coproduct structure. For each Feynman diagram Γ, its coproduct can be written by

$$\Delta(\Gamma) = \Gamma \otimes \mathbb{I} + \mathbb{I} \otimes \Gamma + \sum_{\gamma \subset \Gamma} \gamma \otimes \Gamma/\gamma$$

such that the sum is over all disjoint unions of 1PI superficially divergent proper subgraphs with residue in $R_V \cup R_E$ where the associated amplitudes of their residues need renormalization. Now expand it to the free products of 1PI graphs. □

Remark 3.1.6. *There are different choices for grading structure on the Hopf algebra of Feynman diagrams such as number of vertices, number of internal edges, number of independent loops, ... Grading with the number of internal edges determines finite type property for this Hopf algebra. [19, 22]*

It is remarkable to know that one can reformulate this Hopf algebra by a certain decorated version of the Connes-Kreimer Hopf algebra of rooted trees such that decorations conserve some physical information such as (sub-)divergences (i.e. nested loops) of Feynman diagrams. As an example, the diagram

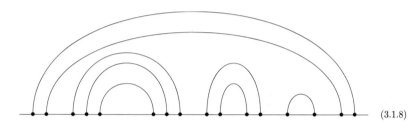

(3.1.8)

can be represented by the labeled rooted tree

(3.1.9)

such that each vertex reports the divergent primitive sub-diagram ⌢ and edges show the locations of sub-diagrams in the main graph. Representation of Feynman diagrams together with overlapping divergences based on rooted trees is also studied. In [53, 71] authors show that how these kinds of nested sub-divergences can be reduced to linear combinations of rooted trees.

The Lie algebra $\mathcal{L}$ gives rise to two representations acting as derivations on $H(\Phi)$. They are

$$< Z^+_{\Gamma_1}, \Gamma_2 > := \Gamma_2 \star \Gamma_1 \qquad (3.1.10)$$

$$< Z^-_{\Gamma_1}, \Gamma_2 > := \sum_i < Z^+_{\Gamma_1}, (\Gamma_2)'_i > (\Gamma_2)''_i \qquad (3.1.11)$$

such that
$$\Delta(\Gamma_2) = \mathbb{I} \otimes \Gamma_2 + \Gamma_2 \otimes \mathbb{I} + \sum_i (\Gamma_2)'_i \otimes (\Gamma_2)''_i. \quad (3.1.12)$$

Remark 3.1.7. *If Γ_2 be a 1PI graph, then for each term in the above sum, there is a unique gluing data G_i that describes how one can reach to the graph Γ_2 by gluing of the components $(\Gamma_2)'_i$ into $(\Gamma_2)''_i$. [16, 37, 57]*

There is also another interesting and useful grading where it can be applied to explain the relation between the Connes-Kreimer Hopf algebra of rooted tress and the Hopf algebra of Feynman diagrams, proof of locality of counterterms and also in the study of Dyson-Schwinger equations.

Definition 3.1.8. *Let $Ker \; \epsilon$ be the augmentation ideal of $H(\Phi)$ and*
$$P : H(\Phi) \longrightarrow ker \; \epsilon, \quad P := id - \mathbb{I}\epsilon$$
be the projection on to this ideal. Define a new map
$$Aug^{(m)} := \underbrace{(P \otimes ... \otimes P)}_{m} \Delta^{m-1} : H(\Phi) \longrightarrow \{ker \; \epsilon\}^{\otimes m}$$
and set
$$H(\Phi)^{(m)} := \frac{Ker \, Aug^{(m+1)}}{Ker \, Aug^{(m)}}, \quad m \geq 1.$$
It is called bidegree.

Lemma 3.1.9. *One can show that*
$$H(\Phi) = \bigoplus_{m \geq 0} H(\Phi)_m = \bigoplus_{m \geq 0} H(\Phi)^{(m)}$$
such that for each $m \geq 1$,
$$H(\Phi)_m \subset \bigoplus_{j=1}^{m} H(\Phi)^{(j)}, \quad H(\Phi)_0 \simeq H(\Phi)^{(0)} \simeq \mathbb{K}.$$
[62, 78]

Remark 3.1.10. *(i) All Feynman graphs that contain (sub-)divergences (i.e. nested loops) belong to the augmentation ideal and it means that $H_{aug}(\Phi) := \bigoplus_{i \geq 1} H(\Phi)_i$ stores quantum information. (ii) For each 1PI graph Γ, one can identify a linear generator δ_Γ and set $H_{lin}(\Phi) := span \; \{\delta_\Gamma\}_\Gamma$. It is observed that $H_{lin}(\Phi) \subset H_{aug}(\Phi)$.*

The grafting operator B^+ is defined on rooted trees but with attention to the decorations one can lift it to the level of Feynman diagrams. For much better understanding of this translation, letting $H_{CK}(\Phi)$ be a labeled version of the Connes-Kreimer Hopf algebra of rooted trees (decorated by primitive 1PI Feynman graphs of the renormalizable theory Φ). By choosing a primitive element γ, the operator B_γ^+ is an homogeneous operator of degree one such that after its application to a forest, it connects the roots in the forest to a new root decorated by γ. As an example, one can see that

$$B^+ \quad \underset{}{} \left(\longrightarrow \right) = \quad (3.1.13)$$

In chapter eight, it will be discussed that the grafting operator determines Hochchild one cocycles of a chain complex connected with the renormalization coproduct. Now bidegree (as the grading factor) and the operator B^+ can provide a decomposition of diagrams which contain primitive components such that it helps us to have a practical instruction for studying Feynman diagrams.

Theorem 3.1.2. *Define a homomorphism* $\Psi : H(\Phi) \longrightarrow H_{CK}(\Phi)$ *given by*

$$\Psi(\Gamma) = \sum_{j=1}^{r} B^+_{\gamma_j, G_{j,i}} [\prod_{i=1}^{k_j} \Psi(\Gamma_{j,i})].$$

One can show that
(i) Ψ *is defined by induction on bidegree,*
(ii) It is a Hopf algebra homomorphism,
(iii) Its image is a closed Hopf subalgebra,
(iv) $G_{j,i}s$ *are the gluing data,*
(v) $B^+_{\gamma_j, G_{j,i}}s$ *are one-cocycles. [16, 37, 57, 78]*

Feynman diagrams of a theory are equipped with physical information. For example external edges have external momenta and because of that it should be reasonable to pay attention to theses physical properties of graphs in the structure of the Hopf algebra. Therefore people prefer to consider this story at two levels: discrete part (i.e. Feynman diagrams without notice to the external structures) for studying the toy model and full structure (i.e. Feynman diagrams together with external data).

Theorem 3.1.3. *For a given renormalizable quantum field theory* Φ,
(i) The discrete Hopf algebra of Feynman diagrams $H_D(\Phi)$ *is made on the free commutative algebra over* $\mathbb{C}$ *generated by pairs* (Γ, w) *such that* Γ *is a 1PI graph and* w *is a monomial with degree agree with the number of external edges of graph.*
(ii) The full Hopf algebra $H_F(\Phi)$ *is made on the symmetric algebra of the linear space of distributions defined by the external structures.*
(iii) $H(\Phi)$ *is isomorphic to* $H_D(\Phi)$ *(in the case: without external structure) or* $H_F(\Phi)$ *(in the case: with external structures). [14, 15, 22, 62]*

It is discussed that how one can arrange Feynman diagrams of given physical theory into a Hopf algebra based on the recognizing of sub-divergences of diagrams. Now it is important to have an explicit understanding from the concept of *"renormalizability"* underlying this Hopf algebra structure and so at the final part of this section we consider this essential concept.

Start with the Largrangian of a given physical theory Φ. We know that each monomial in the Lagrangian corresponds to an amplitude. Letting $\mathcal{A}$ be the set of all amplitudes.

Definition 3.1.11. *The physical theory* Φ *is renormalizable if it has a finite subset* $\mathcal{R}_+ \subset \mathcal{A}$ *as the set of amplitudes which need renormalization.*

For example, it can be easily seen that ϕ^3 in dimension $D \leq 6$ is a renormalizable theory.

Each amplitude $a \in \mathcal{A}$ specifies an integer $n = n(a)$ which gives the number of external edges. Let $\mathcal{M}_a$ be the set of all 1PI graphs contributing to the amplitude a, $|\Gamma|$ be the number of independent loops in Γ and $|sym(\Gamma)|$ be the rank of the automorphism group of the graph.

Definition 3.1.12. *Let ϕ be the Feynman rules character associated to the theory Φ. The Green function related to an amplitude a and the character ϕ is given by*

$$G_\phi^a = 1 \pm \sum_{\Gamma \in \mathcal{M}_a} \alpha^{|\Gamma|} \frac{\phi(\Gamma)}{|sym(\Gamma)|} = 1 \pm \sum_{k \geq 1} \alpha^k \phi(c_k^a)$$

where

$$c_k^a = \sum_{\Gamma \in \mathcal{M}_a, |\Gamma|=k} \frac{\Gamma}{|sym(\Gamma)|}$$

such that the sum is over all 1PI graphs of order k contributing to the amplitude a. The plus sign is taken if $n(a) \geq 3$ and the minus sign for $n(a) = 2$.

Since we are interested to study graphs together with (sub-)divergences, therefore it is not necessary to consider all graphs. Because for instance in ϕ^3 theory in six dimension (as a toy model), one investigates that superficial divergent diagrams are those with the number of external edges ≤ 3 and further *tadpole amplitudes* (i.e. $n(a) = 1$) and *vacuum amplitudes* (i.e. $n(a) = 0$) have vanished Green functions.

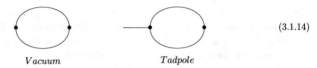

(3.1.14)

Vacuum *Tadpole*

Indeed, it is enough to study Green functions depended on 1PI graphs with 2 or 3 external legs. Because with the help of these graphs, one can build a basis for the Hopf algebra H_{FG}. Now by indicating the structure of the sum

$$\Gamma^a = \mathbb{I} \pm \sum_{\Gamma \in \mathcal{M}_a} \alpha^{|\Gamma|} \frac{\Gamma}{|sym(\Gamma)|} \qquad (3.1.15)$$

in the definition 3.1.12, it is implicitly observed that after the application of the Feynman rules character ϕ on this sum, we will obtain its related Green function and it means that

$$G_\phi^a = \phi(\Gamma^a). \qquad (3.1.16)$$

Lemma 3.1.13. *For a given renormalizable theory Φ, one can decompose the set of amplitudes $\mathcal{A}$ into two disjoint subsets $\mathcal{R}_+$, $\mathcal{R}_-$ such that for each $r \in \mathcal{R}_+$, we have*

$$\Gamma^r = \mathbb{I} \pm \sum_{k \geq 1} \alpha^k B_{k;r}^+ (\Gamma^r Q^{n_r k})$$

where $B_{k;r}^+$ are Hochschild 1-cocycles and Q^{n_r} is a monomial in Γ^r or its inverse. Since the amplitudes from the set $\mathcal{R}_-$ are determined with the knowledge of elements in $\mathcal{R}_+$, therefore for the study of renormalizable theories, it is enough to focus on the elements in $\mathcal{R}_+$. [8, 9, 60, 62, 63, 69]

Recognizing Feynman diagrams together with (sub-)divergences can help us to consider more precisely their related Feynman integrals. There is an important factor (connected with the dimension of theory) to identify divergences in graphs namely, *superficial degree of divergency* ω.

Remark 3.1.14. *Parameter ω has the following properties:*

(i) All Feynman diagrams with the same external structure have the same superficial degree of divergence. It means that

$$res(\Gamma_1) = res(\Gamma_2) \Longrightarrow \omega(\Gamma_1) = \omega(\Gamma_2).$$

(ii) This degree shows the existence of divergency only for a finite number of distinct external structures r in $\mathcal{R}_+$. [37]

Actually, the superficial degree of divergency of a 1PI graph Γ measures the degree of singularity of the integral in the amplitude a_Γ with respect to the integrated variables $q_1, q_2, \ldots$. Under the transformation of momenta $q_i \longmapsto tq_i$ ($t \in \mathbb{R}$), we have $a_\Gamma \longmapsto t^{\omega(\Gamma)} a_\Gamma$. With notice to this factor, a classification of amplitudes of a given renormalizable theory is possible.

Lemma 3.1.15. *The amplitude of a diagram Γ with just one loop*
(i) is convergent if $\omega(\Gamma) < 0$,
(ii) has a logarithmic divergency if $\omega(\Gamma) = 0$,
(iii) has a polynomial divergency if $\omega(\Gamma) > 0$.

The computation of the superficial degree of divergency of a Feynman graph Γ with more than one loop is also possible with starting from 1PI subgraphs with one loop and continue the process by enlarging subgraphs until to reach to the main graph.

Remark 3.1.16. *If we look at to the above procedure, the existence of a self-similar recursive way determines the formal sums Γ^r ($r \in \mathcal{R}_+$) in terms of themselves and the action of suitable maps (i.e. one cocycles) $B_{k;r}^+$.*

For each Feynman diagram Γ, $B_n^+(\Gamma)$ is defined by the insertion of a collection of (sub-)divergences Γ into the identified n−loop primitive graphs. Kreimer could introduce a new measure to translate the combinatorics of Feynman diagrams to the normal analytic picture which physicists familiar with it.

Lemma 3.1.17. *There is a measure μ_+ such that for each Feynman rules character ϕ, it is determined by $\phi(B_n^+(\mathbb{I})) := \int d\mu_+$ where the expression $\phi(B_n^+(\Gamma)) = \int \phi(\Gamma)d\mu_+$ shows that subgraphs become subintegrals under the Feynman rules. [59, 64, 69]*

In summary, it is considered that for each arbitrary renormalizable perturbative physical theory one can introduce a graded connected commutative non-cocommutative Hopf algebra of finite type such that 1PI Feynman graphs play the role of its building blocks. And also, with the help of a decorated version of the Connes-Kreimer Hopf algebra of rooted trees (such that primitive divergent sub-graphs are put in labels), one can display a toy model from this Hopf algebra where it will be useful to do renormalization in a simpler procedure for different theories. This toy model can provide a universal framework in the study of different theories.

3.2 Hopf algebraic formalism

Attempts at eliminating divergences had been started from the birth of quantum field theory. In experimentally calculation of amplitudes, we have encountered two different kinds of divergences:

infra-red (i.e. an amplitude which becomes infinite for vanishingly small values of some momenta) and *ultra-violet* (i.e. an amplitude which becomes infinite for arbitrary large values of momenta in a loop integration). The applied structure in interpretation of ultra-violet divergences will also allow to cancel infra-red divergences. The main idea of renormalization is to correct the original Lagrangian of a quantum field theory by an infinite series of counterterms, labeled with Feynman graphs of the theory. By these counterterms we can disappear ultra-violet divergences. [22, 73, 81]

Feynman diagrams together with some special rules have the ability of presenting all of the possible happenings in a renormalizable theory. For example in a Feynman graph, external edges are the symbols for particles with assigned momenta, vertices show interactions and internal edges for creation and annihilation of virtual particles are applied.

With applying Feynman rules of the given theory, one can associate a Feynman iterated integral to each Feynman diagram. Generally since these integrals have (sub-)divergences, therefore they are complex and ill-defined. This problem is in fact the most conceptional difficulty in quantum field theory such that with the help of various approaches, people try to find suitable solutions for this problem. As a consequence, nowadays renormalization is one understandable powerful technique to consider the behavior of diagrams together with (sub-)divergences. Furthermore, we studied that how one can connect a Hopf algebra structure to a given theory and now it surely very important to know that how this new mathematical structure leads to a new algebraic reformulation from the perturbative renormalization process underlying BPHZ method. It should be remarked that the wide advantage of this point of view (in the study of renormalization) than the Bogoliubov recursive standard process can be observed more clearly, when we want to work on the Feynman diagrams with high loop orders. [16, 31, 53, 62, 64, 65, 109, 110]

Here we are going to have a short overview from renormalization process with respect to this Hopf algebraic modeling. This formalism is discovered by Connes and Kreimer and they could find very closed relation between this well-known physical technique and one important general instruction in mathematics, namely Riemann-Hilbert problem.

There are several mechanisms for renormalization. Connes-Kreimer approach to the perturbative renormalization is certainly a practical reformulation from the BPHZ method on the basis of the Hopf algebra H_{FG} connected with a given renormalizable theory Φ such that Feynman rules can be determined with special characters of this Hopf algebra. One should point out that because of the universal property of the Connes-Kreimer Hopf algebra H_{CK} of rooted tree (with respect to the Hochschild cohomology), the using of a decorated version of this Hopf algebra in computations helps us to find a simplified toy model for studying. [28, 29]

So we should concentrate on the renormalization procedure depended upon one regularization parameter. In this kind, the first stage of renormalization is done by regularization such that with the help of regularization parameter, all divergences appearing in amplitudes can be parameterized to reach to the formally finite expansions together with a subtraction of ill-defined expressions. In the regularization process some non-physical parameters will be created such that this fact could be changed the nature of Feynman rules to algebra homomorphisms from the Hopf algebra H_{FG} to a commutative algebra A where this algebra is characterized by the regularization prescription. Connes and Kreimer proved that minimal subtraction scheme in dimensional regularization can be rewritten based on the Birkhoff decomposition of characters of H_{FG} with values in the algebra A_{dr} of Laurent series with finite pole part such that components of this factorization of very special characters (identified with the Feynman rules of a given theory) provide counterterms and renormalized values. Moreover, one can redefine the renormalization group and its infinitesimal generator (β-function) by using the negative component. In this procedure, one important algebraic property plays an essential role. It is the Rota-Baxter property of the pair (A_{dr}, R_{ms}) such that it reports the multiplicativity of renormalization and it leads to apply the Riemann-Hilbert correspondence in the study of perturbative QFT. [27, 28, 29, 109]

Definition 3.2.1. *Supposing that H be the Hopf algebra of Feynman diagrams of a renormalizable QFT Φ and A be a commutative algebra with respect to the regularization scheme. Set*

$$G(A) := Hom(H, A) = \{\phi : H \longrightarrow A : \phi(xy) = \phi(x)\phi(y), \ \phi(1) = 1\} \subset L(H, A)$$

and consider the convolution product on $G(A)$ such that for each $\phi_1, \phi_2 \in G(A)$, it is given by

$$\phi_1 * \phi_2 := m_A \circ (\phi_1 \otimes \phi_2) \circ \triangle_H.$$

Theorem 3.2.1. *(i) The convolution product $*$ determines a group structure on $G(A)$.*
(ii) For a fixed commutative Hopf algebra H, there is a representable covariant functor $\mathbf{G}$ (represented by H) from the category of unital commutative $\mathbb{K}$−algebras to the category of groups.
(iii) Every covariant representable functor between two above categories is determined by an affine group scheme $\mathbf{G}$. For each commutative algebra A, $\mathbf{G}(A)$ is called affine group scheme. [14, 15, 18, 19, 20, 21, 22]

One can improve this categorical configuration to the Lie algebra level.

Theorem 3.2.2. *For a given affine group scheme $\mathbf{G}$ (viewed as a functor), one can extend it to the covariant functor*

$$\mathfrak{g} = Lie \ \mathbf{G}, \quad A \longmapsto \mathfrak{g}(A)$$

from the category of commutative unital $\mathbb{K}$−algebras to the category of Lie algebras. $\mathfrak{g}(A)$ is the Lie algebra of linear maps $l : H \longrightarrow A$ such that for each $x, y \in H$,

$$l(xy) = l(x)\epsilon(y) + \epsilon(x)l(y)$$

where ϵ is the augmentation of H. [18, 19, 20, 21, 22]

Remark 3.2.2. *Equivalently there is another picture from the elements of $\mathfrak{g}(A)$ given by linear maps $t : H \longrightarrow H$ with the properties*

$$t(xy) = xt(y) + t(x)y, \quad \triangle t = (Id \otimes t) \triangle.$$

In this case, the Lie bracket is defined by the commutator with respect to the composition. [20, 22]

Since characters include Feynman rules of a theory therefore it is important to have enough information about the dual space $L(H, A)$ of all linear maps from H to A. We know that filtration can determine a concept of distance related to Hopf algebras. Loop number of Feynman diagrams induces an increasing filtration on H such that in the dual level, it defines a decreasing filtration on $L(H, A)$. Applying (2.1.18), (2.1.19) and (2.1.20) determine a complete metric $\mathbf{d}_\Phi$ on the space $L(H, A)$ such that it will be seen that this dual space can play an essential role in the study of quantum integrable systems.

It is good place to familiar with another additional algebraic structure namely, Rota-Baxter maps.

Definition 3.2.3. *Let $\mathbb{K}$ be a field with characteristic zero and A an associative unital $\mathbb{K}$−algebra. A $\mathbb{K}$−linear map $R : A \longrightarrow A$ is called Rota-Baxter operator of weight $\lambda \in \mathbb{K}$, if for all $x, y \in A$, it satisfies*

$$R(x)R(y) + \lambda R(xy) = R(R(x)y + xR(y)). \qquad (3.2.1)$$

The pair (A, R) is called Rota-Baxter algebra.

Lemma 3.2.4. *(i) For $\lambda \neq 0$, the standard form*

$$R(x)R(y) + R(xy) = R(R(x)y + xR(y))$$

is given by the transformation $R \longmapsto \lambda^{-1} R$.
(ii) If R is a Rota-Baxter map, then $\widetilde{R} := Id_A - R$ will be a Rota-Baxter map and also $Im\ R$, $Im\ \widetilde{R}$ are subalgebras in A.
(iii) For a given Rota-Baxter algebra (A, R), define a product

$$a \circ_R b := R(a)b + aR(b) - ab$$

on A. $A_R := (A, \circ_R, R)$ is a Rota-Baxter algebra. It is called double Rota-Baxter algebra and one can continue this process to obtain an infinite sequence of doubles.
(iv) The pair (A, R) has a unique Birkhoff decomposition $(R(A), -\widetilde{R}(A)) \subset A \times A$ if and only if it is a Rota-Baxter algebra.
(v) There is a natural way to extend the Rota-Baxter property to the Lie algebra level. For a given Rota-Baxter algebra (A, R), let $(A, [.,.])$ be its Lie algebra with respect to the commutator. One can show that for each $x, y \in A$,

$$[R(x), R(y)] + R([x, y]) = R([R(x), y] + [x, R(y)]). \qquad (3.2.2)$$

Triple $(A, [.,.], R)$ is called Lie Rota-Baxter algebra. [27, 28, 29, 78]

Based on a given Rota-Baxter structure on the algebra A, it is possible to determine a new algebraic structure on the dual space $L(H, A)$ such that it will be applied to reformulate renormalization process.

Theorem 3.2.3. *Suppose regularization and renormalization schemes in a given theory are introduced by Rota-Baxter algebra (A, R). Another Rota-Baxter map Υ can be inherited on $L(H, A)$ given by $\phi \longmapsto \Upsilon(\phi) := R \circ \phi$.*
(i) Triple $\widetilde{\Phi} := (L(H, A), *, \Upsilon)$ is a complete filtered noncommutative associative unital Rota-Baxter $\mathbb{K}$−algebra of weight one.
(ii) One can extend this Rota-Baxter map to the Lie algebra level. [27, 40]

Now consider one particular renormalization method in modern physics namely, minimal subtraction scheme in dimensional regularization. This interesting scheme is identified with the commutative algebra of Laurent series with finite pole part $A_{dr} := \mathbb{C}[[z, z^{-1}]]$ and the renormalization map R_{ms} on A_{dr} where it is given by

$$R_{ms}\Big(\sum_{i \geq -m}^{\infty} c_i z^i\Big) := \sum_{i \geq -m}^{-1} c_i z^i. \qquad (3.2.3)$$

It is an idempotent Rota-Baxter map of weight one. Connes and Kreimer proved that the Bogoliubov's recursive formula for counterterms and renormalized values (depended on determined characters) can be reconstructed with the Birkhoff decomposition on elements of $G(A_{dr})$ and the map R_{ms} [14, 15, 22].

Theorem 3.2.4. *For a given renormalizable physical theory Φ underlying the BPHZ method with the related Hopf algebra $H_D(\Phi)$ (for the discrete level) and $H_F(\Phi)$ (for the full level),*
(i) The groups of diffeographisms $Diffg(\Phi)_D$ (for the discrete level) and $Diffg(\Phi)_F$ (for the full level) are the pro-unipotent affine group schemes of the Hopf algebras $H_D(\Phi)$ and $H_F(\Phi)$. The relation between these two groups is determined by a semidirect product as follows

$$Diffg(\Phi)_F = Diffg(\Phi)_{ab} \rtimes Diffg(\Phi)_D.$$

(ii) Set $H = H(\Phi)$. The graded dual Hopf algebra H^* contains all finite linear combinations of homogeneous linear maps on H. If $L := Lie\ PrimH^*$, then there is a canonical isomorphism of Hopf algebras between H and the graded dual of the universal enveloping algebra of L and moreover, $L = Lie\ G(A_{dr})$. [19, 22]

The phrase "*diffeographism*" is motivated from this fact that the space Diffg(Φ) acts on coupling constants of the theory through formal diffeomorphisms tangent to the identity. These diffeographisms together with Birkhoff factorization provide enough information to calculate counterterms and renormalized values.

Theorem 3.2.5. *For the dimensionally regularized Feynman rules character $\phi \in G(A_{dr})$, there is a unique pair (ϕ_-, ϕ_+) of characters in $G(A_{dr})$ such that*

$$\phi = \phi_-^{-1} * \phi_+.$$

It determines an algebraic Birkhoff decomposition for the chosen character. [20, 27, 40]

Based on the Riemann-Hilbert correspondence as a motivation, Connes and Kreimer proved that physical parameters of a given theory can reformulate with the characters ϕ_- and ϕ_+. It was the first bridge between Birkhoff decomposition and theory of quantum fields.

Theorem 3.2.6. *For the dimensionally regularized Feynman rules character ϕ in $G(A_{dr})$,*
(i) Its related Birkhoff components are determined by

$$\phi_-(\Gamma) = e_{A_{dr}} \circ \epsilon_H(\Gamma) - \Upsilon(\phi_- * (\phi - e_{A_{dr}} \circ \epsilon_H))(\Gamma)$$
$$= -R_{ms}(\phi(\Gamma) + \sum_{\gamma \subset \Gamma} \phi_-(\gamma)\phi(\frac{\Gamma}{\gamma})),$$
$$\phi_+(\Gamma) = e_{A_{dr}} \circ \epsilon_H(\Gamma) - \widetilde{\Upsilon}(\phi_+ * (\phi^{-1} - e_{A_{dr}} \circ \epsilon_H))(\Gamma)$$
$$= \phi(\Gamma) + \phi_-(\Gamma) + \sum_{\gamma \subset \Gamma} \phi_-(\gamma)\phi(\frac{\Gamma}{\gamma}),$$

such that $\Gamma \in ker\ \epsilon_H$ and the sum is over all disjoint unions of superficially divergent 1PI proper subgraphs.

(ii) BPHZ renormalization (i.e. unrenormalized regularized value, counterterms, renormalized values) can be rewritten algebraically by

$$\phi(\Gamma) = U_\mu^z(\Gamma), \quad \phi_-(\Gamma) = c(\Gamma), \quad \phi_+(\Gamma) = rv(\Gamma).$$

[14, 15, 20, 27, 40]

To remove divergences step by step in a perturbative expansion of Feynman diagrams is the main idea of renormalization and in this language a theory is renormalizable, if the disappearing of all divergences be possible by such a finite recursive procedure. With attention to the algebraic reformulation of the BPHZ prescription, in the next step of this section, we consider the structure of the renormalization group and its related infinitesimal character. We observe that how geometric concepts (such as loop space) would be entered in the story to provide a complete algebro-geometric description from physical information.

Let $U_\mu^z(\Gamma(p_1,...p_N))$ be a regularized unrenormalized value. It determines a loop γ on the infinitesimal punctured disk Δ^* (connected with the regularization parameter) around the $z = 0$ with values in the group of diffeographisms such that this loop has a Birkhoff factorization

$$\gamma(z) = \gamma_-(z)^{-1}\gamma_+(z). \tag{3.2.4}$$

where $\gamma_-(z)$ (holomorphic in $\mathbb{P}(\mathbb{C}) - \{0\}$) gives the counterterm and $\gamma_+(z)$ (regular at $z = 0$) determines the renormalized value. Since $U_\mu^z(\Gamma(p_1,...p_N))$ depends on the mass parameter μ, its related loop should have a dependence on this parameter. In summary,

$$U_\mu^z(\Gamma(p_1,...p_N)) \Longleftrightarrow \gamma_\mu : \Delta^* \longrightarrow Diffg(\Phi). \tag{3.2.5}$$

Now one can see that the space $Loop\ (Diffg(\Phi), \mu)$ contains physical information of the theory Φ.

Theorem 3.2.7. *Let $G(\mathbb{C})$ be the pro-unipotent complex Lie group associated to the positively graded connected commutative finite type Hopf algebra H of the renormalizable theory Φ underlying the minimal subtraction scheme in dimensional regularization. Suppose $\gamma_\mu(z)$ be a loop with values in $G(\mathbb{C})$ that encodes $U_\mu^z(\Gamma(p_1,...,p_n))$. It has a unique Birkhoff decomposition*

$$\gamma_\mu(z) = \gamma_{\mu-}(z)^{-1}\gamma_{\mu+}(z)$$

such that
(i) $\frac{\partial}{\partial \mu}\gamma_{\mu-}(z) = 0$,
(ii) For each $\phi \in H_n^*$, $t \in \mathbb{C} : \theta_t(\phi) = e^{nt}\phi$ is a $1-$parameter group of automorphisms on $G(\mathbb{C})$,
(iii) $\gamma_{e^t\mu}(z) = \theta_{tz}(\gamma_\mu(z))$,
(iv) The limit

$$F_t = lim_{z \longrightarrow 0}\gamma_-(z)\theta_{tz}(\gamma_-(z)^{-1})$$

exists and it denotes a $1-$parameter subgroup F_t of $G(\mathbb{C})$. It means that for each s, t,

$$F_{s+t} = F_s * F_t.$$

(v) For each Feynman diagram Γ, $F_t(\Gamma)$ is a polynomial in t.
(vi) $\forall t \in \mathbb{R} : \gamma_{e^t\mu+}(0) = F_t\gamma_{\mu+}(0)$. [14, 15, 19, 20, 22]

Definition 3.2.5. *The 1−parameter subgroup $\{F_t\}_t$ of $G(\mathbb{C})$ identifies the renormalization group of the theory such that its infinitesimal generator*

$$\beta = \frac{d}{dt}|_{t=0} F_t$$

determines the β−function of the theory.

Remark 3.2.6. *$G(\mathbb{C})$ is a topological group with the topology of pointwise convergence and it means that for each sequence $\{\gamma_n\}_n$ of loops with values in $G(\mathbb{C})$ and for each Feynman diagram $\Gamma \in H$,*

$$\gamma_n \longrightarrow \gamma \iff \gamma_n(\Gamma) \longrightarrow \gamma(\Gamma).$$

Letting $\mathfrak{g}(\mathbb{C})$ be the Lie algebra of diffeographisms of Φ. We know that it contains all of the linear maps $Z : H \longrightarrow A$ satisfying in the Libniz's law namely, derivations such that its Lie bracket is given by the commutator with respect to the convolution product. There is a bijection between this Lie algebra and its corresponding Lie group given by the exponential map. One can expand the Lie algebra $\mathfrak{g}(\mathbb{C})$ with an additional generator Z_0 such that for each $Z \in \mathfrak{g}(\mathbb{C})$,

$$[Z_0, Z] = Y(Z) \tag{3.2.6}$$

where Y is the grading operator.

Lemma 3.2.7. *(i) $\frac{d\theta_t}{dt}|_{t=0} = Y$.*
(ii) For each $\phi \in H^*$ and Feynman diagram $\Gamma \in H$, we have

$$<\theta_t(\phi), \Gamma> = <\phi, \theta_t(\Gamma)>.$$

It means that $\theta_t = Ad_{exp^*(tZ_0)}$. [19, 20, 28]

In a more practical point of view and for the simplicity in computations, one can have a scattering type formula for components of the dimensionally regularized Feynman rules character ϕ. It means that

Lemma 3.2.8. *(i) $F_t = lim_{z \to 0} \phi_- \theta_{tz} \phi_-^{-1}$,*
(ii) $\beta(\phi) = \phi_\pm * Y(\phi_\pm^{-1}) = \phi_\pm * [Z_0, \phi_\pm^{-1}]$,
(iii) $exp^*(t(\beta(\phi) + Z_0)) * exp^*(-tZ_0) = \phi_\pm * \theta_t(\phi_\pm^{-1}) \longrightarrow^{t \to \infty} \phi_\pm$. [28, 29, 93]

Finally, let us recalculate physical information to emphasize more the role of the Connes-Kreimer Hopf algebra in this algebraic formalism. Indeed, this machinery works with the antipode map [64]. If ϕ be the Feynman rules character, then consider the undeformed character $\phi \circ S$ and deform this character by the renormalization map. So for each Feynman diagram Γ, the BPHZ renormalization can be summarized in the equations

$$S^\phi_{R_{ms}}(\Gamma) = -R_{ms}(\phi(\Gamma)) - R_{ms}(\sum_{\gamma \subset \Gamma} S^\phi_{R_{ms}}(\gamma) \phi(\frac{\Gamma}{\gamma})), \tag{3.2.7}$$

$$\Gamma \longmapsto S^\phi_{R_{ms}} * \phi(\Gamma). \tag{3.2.8}$$

Because it is easy to see that

$$S^\phi_{R_{ms}} * \phi(\Gamma) = \overline{R}(\Gamma) + S^\phi_{R_{ms}}(\Gamma) \tag{3.2.9}$$

such that the Bogoliubov operation $\overline{R}$ is given by

$$\overline{R}(\Gamma) = U_\mu^z(\Gamma) + \sum_{\gamma \subset \Gamma} c(\gamma) U_\mu^z(\frac{\Gamma}{\gamma}) = \phi(\Gamma) + \sum_{\gamma \subset \Gamma} S_{R_{ms}}^\phi(\gamma) \phi(\frac{\Gamma}{\gamma}). \quad (3.2.10)$$

It means that $S_{R_{ms}}^\phi(\Gamma)$ and $S_{R_{ms}}^\phi * \phi(\Gamma)$ are counterterm and renormalized value depended on the Feynman diagram Γ.

In summary, Connes-Kreimer perturbative renormalization introduces a new algebraic interpretation to calculate the renormalization group based on the loop space of characters and Birkhoff factorization. The geometric nature of this procedure will be shown more, when we consider the renormalization bundle and its related flat equi-singular connections (formulated by Connes and Marcolli). Moreover, we know that the minimal subtraction scheme in dimensional regularization indicates a Rota-Baxter algebra such that it provides Birkhoff decomposition for characters. As we shall see later that how this algebraic property helps us to indicate a new approach to consider theory of integrable systems in renormalizable quantum field theories such that renormalization has a central role.

Chapter 4

Integrable Renormalization

It was shown that with the help of Hopf algebra of renormalization on Feynman diagrams of a renormalizable QFT and based on components of the Birkhoff decomposition of some particular elements in the loop space on diffeographisms, one can determine counterterms, renormalized values, the Connes-Kreimer renormalization group and its infinitesimal generator. Indeed, the identification of renormalization with the Riemann-Hilbert problem provides a new conceptual interpretation of physical information.

With applying Atkinson theorem, one can show that the existence and the uniqueness of this factorization is a direct consequence of an important and interesting concept in physics namely, multiplicativity of renormalization which is prescribed mathematically with the Rota-Baxter condition of the chosen renormalization scheme. On the other hand and in a roughly speaking, one can find the importance of this class of equations in the study of theory of integrable systems specially, (modified) Yang-Baxter equations. This fact reports obviously a pure algebraic configuration of renormalization where it indeed seems wise to introduce a coherent ideology for considering quantum integrable systems with respect to the Connes-Kreimer theory. [26, 28, 29, 62, 68]

With attention to the theory of integrable systems in classical level [11, 23, 66, 67, 68, 95], this (rigorous) algebraic machinery for the description of renormalizable QFTs and the power of noncommutative differential forms [38], we are going to find a new family of Hamiltonian systems which arise from the Connes-Kreimer approach to perturbative renormalization and moreover, we show that how integrability condition can be determined naturally based on Poisson brackets related to Rota-Baxter type algebras. The beauty of this new viewpoint to integrable systems will be identified, when we consider the minimal subtraction underlying the dimensional regularization (as the renormalization mechanism) and then study possible relation between introduced motion integrals and renormalization group. After that, based on the Rosenberg framework, we familiar with other group of quantum Poisson brackets and their related motion integrals. We close this chapter with introducing a new family of fixed point equations modified with motion integral condition and Bogoliubov character. In summary, it seems favorable to report about the appearance of a glimpse of one general relation between the theory of Rota-Baxter type algebras and the Riemann-Hilbert problem underlying quantum field theory. [98]

4.1 Integrable systems: from finite dimension (geometric approach) to infinite dimension (algebraic approach)

Let V be a m-dimensional real vector space and $\omega : V \times V \longrightarrow \mathbb{R}$ a skew-symmetric bilinear map.

Definition 4.1.1. *For a map $\tilde{\omega} : V \longrightarrow V^*$ given by*

$$\tilde{\omega}(v)(u) := \omega(v, u),$$

ω is called a symplectic form, if $\tilde{\omega}$ be a bijective map. The pair (V, ω) is called symplectic vector space and it is clear that each symplectic vector space has even dimension.

Definition 4.1.2. *A differential 2-form ω on a manifold M is called symplectic, if it is closed and for each $p \in M$, ω_p is a symplectic form on $T_p M$.*

Definition 4.1.3. *(i) A complex structure on V is a linear map $J : V \longrightarrow V$ such that $J^2 = -Id_V$. (V, J) is called a complex vector space.*
(ii) An almost complex structure on a manifold M is a smooth field of complex structures on the tangent spaces and it means that for each $x \in M$,

$$x \longmapsto J_x : T_x M \longrightarrow T_x M : Linear, \quad J_x^2 = -Id.$$

The pair (M, J) is called an almost complex manifold.
(iii) An almost complex structure J is called integrable, if J is induced by a structure of complex manifold on M.

There is a well known operator to characterize integrable almost complex structures on manifolds. It will be shown that this map is in fact a starting key for us to consider quantum integrable systems underlying an algebraic formalism.

Definition 4.1.4. *For the almost complex manifold (M, J), the Nijenhuis tensor is defined by*

$$NT(v, w) := [Jv, Jw] - J[v, Jw] - J[Jv, w] - [v, w]$$

such that v, w are vector fields on M and for each $f \in C^\infty(M)$,

$$[v, w].f = v.(w.f) - w.(v.f)$$

where $v.f = df(v)$.

Theorem 4.1.1. *For the almost complex manifold (M, J), the following facts are equivalent:*
(i) M is a complex manifold,
(ii) J is integrable,
(iii) $NT \equiv 0$,
(iv) $d = \partial + \overline{\partial}$. [5, 11]

Definition 4.1.4 and theorem 4.1.1 show that if J is an integrable structure, then for each $x \in M$ and $v, w \in T_x M$, we have

$$[J_x v, J_x w] = J_x [v, J_x w] + J_x [J_x v, w] + [v, w]. \tag{4.1.1}$$

Equations of motion are the results of variational problems in classical mechanics and in a system with n particles in $\mathbb{R}^n$, all of the physical trajectories follow from the Newton's second law. In these paths the mean value of the difference between kinetics and potential energies is minimum.

Lemma 4.1.5. *Let (M,ω) be a symplectic manifold and $H: M \longrightarrow \mathbb{R}$ a smooth function. There is a unique vector field X_H on M such that $i_{X_H}\omega = dH$. One can identify a 1-parameter family of diffeomorphisms $\rho_t : M \longrightarrow M$ such that*

$$\rho_0 = Id_M, \quad \frac{d\rho_t}{dt} \circ \rho_t^{-1} = X_H, \quad \rho_t^*\omega = \omega.$$

[5, 11, 66]

This lemma shows that each smooth function $H: M \longrightarrow \mathbb{R}$ determines a family of *symplectomorphisms*. The function H is called *Hamiltonian function* such that the vector field X_H is its corresponding *Hamiltonian vector field*.

Definition 4.1.6. *The triple (M,ω,H) is called a classical Hamiltonian system.*

On the Euclidean space $\mathbb{R}^{2n}$ with the local coordinate $(q_1,...,q_n,p_1,...,p_n)$ and the canonical symplectic form $\omega = \sum dq_i \wedge dp_i$, the curve $\alpha_t = (q(t),p(t))$ is an *integral curve* of the vector field X_H if and only if it determines the Hamiltonian equations of motion

$$\frac{dq_i}{dt} = \frac{\partial H}{\partial p_i}, \quad \frac{dp_i}{dt} = -\frac{\partial H}{\partial q_i} \quad (4.1.2)$$

on $\mathbb{R}^{2n}$.

Theorem 4.1.2. *The Newton's second law on the configuration space $\mathbb{R}^n$ is equivalent to the Hamiltonian equations of motion on the phase space $\mathbb{R}^{2n}$. [5, 11, 66]*

For a given symplectic manifold M and $f,g \in C^\infty(M)$, letting X_f, X_g be the Hamiltonian vector fields with respect to these functions. From this class of vector fields, one expects a Poisson bracket on $C^\infty(M)$ given by

$$\{f,g\} := \omega(X_f, X_g). \quad (4.1.3)$$

The pair $(C^\infty(M), \{.,.\})$ is called *Poisson algebra related to the configuration space M* and from now M is called a *Poisson manifold*.

Remark 4.1.7. *Dual space of a Lie algebra is one useful example of a Poisson manifold but generally, these Poisson manifolds are not symplectic.*

Lemma 4.1.8. *For any $H \in C^\infty(M)$, its related Hamiltonian vector field which acts on the elements of $C^\infty(M)$ (by the Poisson bracket) can be rewritten with the equation*

$$X_H f = \{H, f\}$$

such that for each $x \in M$, the vectors $X_H(x)$ span a linear subspace in T_xM. [5, 11, 66]

We know that Hamiltonian vector fields are tangent to symplectic leaves and it means that the Hamiltonian flows enable to preserve each leaf separately. So for a given Hamiltonian system (M,ω,H), equation

$$\{f,H\} = 0 \quad (4.1.4)$$

45

is equivalent with this fact that f is constant along the integral curves of X_H. The function f is called *integral of motion*. [11, 95]

Definition 4.1.9. *The Hamiltonian system* (M, ω, H) *is called integrable, if there exist* $n = \frac{1}{2} dim M$ *independent integrals of motion* $f_1 = H, f_2, ..., f_n$ *such that* $\{f_i, f_j\} = 0$. *[5, 11]*

Suppose $\mathfrak{g}$ be a Lie algebra with the dual space $\mathfrak{g}^\star$ and $P(\mathfrak{g}^\star)$ be the space of polynomials on the dual space. Elements of the Lie algebra can determine a bracket on $P(\mathfrak{g}^\star)$ such that for each $g_1, g_2 \in \mathfrak{g}$ and $h^\star \in \mathfrak{g}^\star$, it is given by

$$\{g_1, g_2\}(h^\star) := h^\star([g_1, g_2]). \tag{4.1.5}$$

Since $P(\mathfrak{g}^\star)$ is dense in $C^\infty(\mathfrak{g}^\star)$, one can expand canonically this bracket to obtain a *Lie Poisson bracket* such that for each $s_1, s_2 \in C^\infty(\mathfrak{g}^\star)$, we have

$$\{s_1, s_2\}(h^\star) = h^\star([ds_1(h^\star), ds_2(h^\star)]). \tag{4.1.6}$$

Theorem 4.1.3. *For a given Lie group* G *with the Lie algebra* $\mathfrak{g}$, *the symplectic leaves of (4.1.6) are* $G-orbits$ *in* $\mathfrak{g}^\star$. *[5, 66, 95]*

One can apply automorphisms on a Lie algebra to introduce other Lie Poisson brackets. This method provides a new favorable point of view to study infinite dimensional integrable systems.

Definition 4.1.10. *For a fixed Lie algebra* $\mathfrak{g}$, *an endomorphism* $R \in End(\mathfrak{g})$ *is called R-matrix, if for each* $g_1, g_2 \in \mathfrak{g}$, *the bracket*

$$[g_1, g_2]_R = \frac{1}{2}([R(g_1), g_2] + [g_1, R(g_2)])$$

satisfies in the Jacobi identity and it means that $[., .]_R$ *is a Lie bracket.*

Yang-Baxter equation is enough to introduce a new class of Poisson brackets on $\mathfrak{g}^\star$ which arise from this kind of Lie brackets.

Theorem 4.1.4. *Let* $R \in End(\mathfrak{g})$. *For each* $g_1, g_2 \in \mathfrak{g}$, *set*

$$B_R(g_1, g_2) := [R(g_1), R(g_2)] - R([R(g_1), g_2] + [g_1, R(g_2)]).$$

The R−bracket defined in 4.1.10 follows the Jacobi identity if and only if for each $g_1, g_2, g_3 \in \mathfrak{g}$, *we have*

$$[B_R(g_1, g_2), g_3] + [B_R(g_2, g_3), g_1] + [B_R(g_3, g_1), g_2] = 0.$$

This condition is called classical Yang-Baxter equation. The simplest sufficient condition is given by $B_R(g_1, g_2) = 0$. *[66, 67, 68, 95]*

Yang-Baxter equations make possible to consider integrable systems in an algebraic framework such that this ability together with the Hopf algebra of Feynman diagrams help us to study quantum Hamiltonian systems and integrability condition depended on the algebraic renormalization. The critical point in this procedure can be summarized in reinterpretation of the renormalization group on the basis of Rota-Baxter type algebras. We show that how one can find a new family of symplectic structures where its existence is strongly connected with the choosing of renormalization prescription.

4.2 Rota-Baxter type algebras

In this part we dwell a moment on a special class of Rota-Baxter type algebras namely, Nijenhuis algebras to discuss that how one can apply these maps to deform the initial product. It gives a family of new associative algebras together with related compatible Lie brackets.

For a linear map $N : A \longrightarrow A$, define a new product on A. It is given by

$$(x,y) \longmapsto x \circ_N y := N(x)y + xN(y) - N(xy). \tag{4.2.1}$$

Remark 4.2.1. *If e be the unit of (A, m) and $N(e) = e$ then $(A, \circ_N)$ has the same unit.*

Lemma 4.2.2. *(i) $(A, \circ_N)$ is an algebra.*
(ii) The product $\circ_N$ is associative if and only if for each $x, y \in A$,

$$T_N(x,y) := N(x \circ_N y) - N(x)N(y)$$

be a Hochschild 2-cocycle of the algebra A (with respect to the Hochschild coboundary operator b connected with the product of A). It means that for each $x, y, z \in A$,

$$bT_N(x,y,z) := xT_N(y,z) - T_N(xy,z) + T_N(x,yz) - T_N(x,y)z = 0.$$

(iii) N is a derivation in the original algebra if and only if N is a 1-cocycle with respect to the coboundary operator b. In this case, the new product $\circ_N$ is trivial. [12, 24]

Definition 4.2.3. *The linear map N is called Nijenhuis tensor supported by $\circ_N$, if*

$$N(x \circ_N y) = N(x)N(y). \tag{4.2.2}$$

The pair (A, N) is called Nijenhuis algebra.

Remark 4.2.4. *(i) If N is a Nijenhuis tensor, then $\circ_N$ is an associative product on A and also for each $\lambda \in \mathbb{K}$, $m + \lambda \circ_N$ is an associative product on A (such that m is the original product of A).*
(ii) For a given Rota-Baxter algebra (A, R) with the idempotent Rota-Baxter map R and for each $\lambda \in \mathbb{K}$, the operator $N_\lambda := R - \lambda \widetilde{R}$ has Nijenhuis property and it means that (A, N_λ) is a Nijenhuis algebra. [27, 28, 29]

The classical Yang-Baxter equation is essentially governed by the extension of this kind of operators to the Lie algebra level.

Definition 4.2.5. *A Nijenhuis tensor for the Lie algebra $(A, [., .])$ is a linear map $N : A \longrightarrow A$ such that for each $x, y \in A$,*

$$N([x,y]_N) = [N(x), N(y)]$$

where

$$[x,y]_N := [N(x), y] + [x, N(y)] - N([x,y]).$$

Remark 4.2.6. *The compatibility of this Lie bracket is strongly related to the Nijenhuis property of N and it is easy to see that in this case*

$$[N(x), N(y)] = N([N(x), y] + N([x, N(y)]) - N^2([x,y]).$$

The triple $(A, [., .], N)$ is called Nijenhuis Lie algebra. [24]

47

Theorem 4.2.1. *Let $(A, \circ_N)$ be the associative algebra with respect to the Nijenhuis tensor N on A. N is a Nijenhuis tensor for the Lie algebra $(A, [.,.])$ (i.e. $[.,.]$ is the commutator with respect to the product m) and for each $x, y \in A$, we will have*

$$[x,y]_N = x \circ_N y - y \circ_N x.$$

[12, 24, 27, 28, 29]

Because of the importance of Nijenhuis algebras, we are going to mention an algorithmic instruction for constructing this type of algebras from each arbitrary commutative algebra. Let (A, m) be a commutative $\mathbb{K}$-algebra with the tensor algebra $T(A) := \oplus_{n \geq 0} A^{\otimes n}$. Elements of A play the role of *letters* and generators $U = a_1 \otimes ... \otimes a_n$ in $A^{\otimes n}$ (where $a_i \in A$) are identified with words $a_1...a_n$.

Lemma 4.2.7. *For each $a, b \in A$, $U \in A^{\otimes n}$, $V \in A^{\otimes m}$ and $\lambda \in \mathbb{K}$, one can define recursively an associative commutative quasi-shuffle product on $T(A)$ given by*

$$aU \star bV := a(U \star bV) + b(aU \star V) - \lambda m(a,b)(U \star V)$$

such that the empty word plays the role of its unit. [25, 26, 42]

The product given by the lemma 4.2.7 determines another kind of shuffle product.

Lemma 4.2.8. *An augmented quasi-shuffle product on the augmented tensor module $\overline{T}(A) := \oplus_{n > 0} A^{\otimes n}$ is given by*

$$aU \odot bV := m(a,b)(U \star V).$$

It is an associative commutative product such that the unit e of the algebra A is the unit for this product. [25, 26, 42]

One can define a modified version of the above products.

Lemma 4.2.9. *The modified quasi-shuffle product on $T(A)$ is defined by*

$$aU \ominus bV := a(U \ominus bV) + b(aU \ominus V) - em(a,b)(U \ominus V)$$

such that the empty word is the unit for this commutative associative product. Its augmented version namely, augmented modified quasi-shuffle product on $\overline{T}(A)$ is defined by

$$aU \oslash bV := m(a,b)(U \ominus V)$$

such that the unit e of A is the unit for this new product. [25, 26, 42]

These products can be applied to make a special family of Rota-Baxter and Nijenhuis algebras with the universal property.

Theorem 4.2.2. *For a unital commutative associative $\mathbb{K}$-algebra A with the unit e, let B_e^+ be a linear operator on $\overline{T}(A)$ given by*

$$B_e^+(a_1...a_n) := ea_1...a_n.$$

(i) $(T(A), \star, \lambda, B_e^+)$ is a RB algebra of weight λ.

(ii) $(T(A), \ominus, B_e^+)$ *is a Nijenhuis algebra.*

(iii) $(\overline{T}(A), \odot, \lambda, B_e^+)$ *is the universal RB algebra of weight λ generated by A.*

(iv) $(\overline{T}(A), \oslash, B_e^+)$ *is the universal Nijenhuis algebra generated by A. [25]*

Remark 4.2.10. *For instance, universal Nijenhuis algebra (generated by A) means that for any Nijenhuis algebra B and algebra homomorphism $f : A \longrightarrow B$, there exists a unique Nijenhuis homomorphism $\widetilde{f} : \overline{T}(A) \longrightarrow B$ such that $\widetilde{f} \circ j_A = f$.*

Everything is prepared to introduce a new class of quantum Hamiltonian systems as the consequence of the Connes-Kreimer algebraic framework to perturbative theory. It will be shown that how the Hopf algebraic renormalization group can give us some examples of integrable systems from this class of Hamiltonian systems.

4.3 Theory of quantum integrable systems

In this part we want to focus on the algebraic basis of the Connes-Kreimer theory namely, the Rota-Baxter property induced from renormalization to improve theory of integrable systems to the level of renormalizable physical theories. With the help of noncommutative differential forms (associated with Hamiltonian derivations) and with attention to the chosen renormalization method (i.e. regularization algebra and renormalization map), we are going to introduce a new family of Hamiltonian systems based on the Connes-Kreimer Hopf algebra of Feynman diagrams. It is discussed that how integrability condition on these systems are strongly connected with the perturbative renormalization process.

Renormalization prescription makes possible two different types of deformations. In one class, we consider deformed algebras which deformation process is performed by an idempotent renormalization map and in another class, with respect to regularization scheme (independent of the renormalization map), we turn to Ebrahimi-Fard's aspect in defining the universal Nijenhuis algebra (as the kind of deformation method) and then we will deform the initial algebra by this universal Nijenhuis tensor. Finally, with working on differential forms of these deformed algebras, we will illustrate (as the conclusion) Hamiltonian systems and also integrability condition.

Roughly speaking, Connes-Kreimer Hopf algebra and regularization algebra determines a new noncommutative algebra such that with attention to the renormalization scheme, one can provide new deformed algebras. Then with working on the noncommutative differential calculus with respect to these mentioned deformed algebras, we can obtain symplectic structures and so Hamiltonian systems. Because of applying this noncommutative differential formalism on the deformed algebras depended upon the renormalization process, it does make sense to use the phrase "*integrable renormalization*" for this new captured approach to quantum integrable systems.

First Class. Suppose Φ be a renormalizable QFT with the associated Hopf algebra of Feynman diagrams H and letting the perturbative renormalization is performed with the idempotent renormalization map R and the regularization scheme A. In summary, we denote its algebraic reformulation with

$$\widetilde{\Phi} = (L(H, A), *, \Upsilon) \qquad (4.3.1)$$

such that the idempotent Rota-Baxter map Υ on $L(H, A)$ is given by R. It is easy to show that for each $\lambda \in \mathbb{K}$, the operator $\Upsilon_\lambda := \Upsilon - \lambda \widetilde{\Upsilon}$ (where $\widetilde{\Upsilon} := Id - \Upsilon$) has Nijenhuis property.

Definition 4.3.1. *By the formula (4.2.1), for each λ a new product $\circ_\lambda$ on $L(H, A)$ can be defined such that for each $\phi_1, \phi_2 \in L(H, A)$,*

$$\phi_1 \circ_\lambda \phi_2 := \Upsilon_\lambda(\phi_1) * \phi_2 + \phi_1 * \Upsilon_\lambda(\phi_2) - \Upsilon_\lambda(\phi_1 * \phi_2).$$

Remark 4.3.2. *(i) One can show that*

$$\Upsilon_\lambda(\phi_1 \circ_\lambda \phi_2) = \Upsilon_\lambda(\phi_1) * \Upsilon_\lambda(\phi_2).$$

(ii) Lemma 4.2.2 shows that $\circ_\lambda$ is an associative product and one can identify the following compatible Lie bracket

$$[\phi_1, \phi_2]_\lambda := [\Upsilon_\lambda(\phi_1), \phi_2] + [\phi_1, \Upsilon_\lambda(\phi_2)] - \Upsilon_\lambda([\phi_1, \phi_2])$$

such that definition 4.3.1 provides that

$$[\phi_1, \phi_2]_\lambda = \phi_1 \circ_\lambda \phi_2 - \phi_2 \circ_\lambda \phi_1.$$

It means that one can extend the Nijenhuis property of Υ_λ to the Lie algebra level (with the commutator with respect to the product $*$).

Definition 4.3.3. *The new spectral information $(\widetilde{\Phi}_\lambda, [.,.]_\lambda) := (L(H, A), \circ_\lambda, \Upsilon_\lambda, [.,.]_\lambda)$ is called λ-information based on the theory Φ and with respect to the Nijenhuis map Υ_λ.*

Second Class. Generally, when we study the algebraic renormalization methods, the renormalization map might not have idempotent property and therefore we should apply another technique to receive Nijenhuis tensors. In this situation one can focus on a commutative algebra (which reflects the regularization scheme) and apply the universal Nijenhuis tensor based on this algebra.

Let Φ be a renormalizable theory with the related Hopf algebra H such that the regularization scheme is given by the associative commutative unital algebra A. Theorem 4.2.2 shows that $(\overline{T}(A), \oslash, B_e^+)$ is the universal Nijenhuis algebra based on A and therefore one can define a Nijenhuis map Υ_e^+ on $L(H, \overline{T}(A))$ given by

$$\Upsilon_e^+(\psi) := B_e^+ \circ \psi. \tag{4.3.2}$$

Definition 4.3.4. *With help of the operator Υ_e^+, a new product $\circ_u$ on $L(H, \overline{T}(A))$ is introduced such that for each $\psi_1, \psi_2 \in L(H, \overline{T}(A))$,*

$$\psi_1 \circ_u \psi_2 := \Upsilon_e^+(\psi_1) *_\oslash \psi_2 + \psi_1 *_\oslash \Upsilon_e^+(\psi_2) - \Upsilon_e^+(\psi_1 *_\oslash \psi_2),$$

where

$$\psi_1 *_\oslash \psi_2 := \oslash(\psi_1 \otimes \psi_2) \circ \Delta_H.$$

Remark 4.3.5. *(i) The Nijenhuis property of Υ_e^+ provides that*

$$\Upsilon_e^+(\psi_1 \circ_u \psi_2) = \Upsilon_e^+(\psi_1) *_\oslash \Upsilon_e^+(\psi_2),$$

such that it supports the associativity of the product $\circ_u$.
(ii) A compatible Lie bracket $[.,.]_u$ can be defined on $L(H, \overline{T}(A))$ by

$$[\psi_1, \psi_2]_u := [\Upsilon_e^+(\psi_1), \psi_2] + [\psi_1, \Upsilon_e^+(\psi_2)] - \Upsilon_e^+([\psi_1, \psi_2]).$$

Moreover we have
$$[\psi_1, \psi_2]_u = \psi_1 \circ_u \psi_2 - \psi_2 \circ_u \psi_1.$$
It means that Nijenhuis property of Υ_e^+ can be extended to the Lie algebra level (with the commutator with respect to the product $*_\oslash$).

Definition 4.3.6. *The spectral information* $(\widetilde{\Phi}_u, [.,.]_u) := (L(H, \overline{T}(A)), \circ_u, \Upsilon_e^+, [.,.]_u)$ *is called* $u-$*information based on the theory* Φ *and with respect to the Nijenhuis map* Υ_e^+.

Here we introduce a certain family of symplectic structures related to this algebraic preparation from renormalizable physical theories and for this goal we need theory of noncommutative differential calculus over an algebra based on the space of its derivations.

Definition 4.3.7. *Let C be an associative unital algebra over the field $\mathbb{K}$ (with characteristic zero) with the center $Z(C)$. A derivation $\theta : C \longrightarrow C$ is an infinitesimal automorphism of C such that it is a linear map satisfying the Leibniz rule where if it could have been exponentiated, then the map $\exp \theta$ will be an automorphism of C.*

Remark 4.3.8. *(i) Geometrically, θ is a vector field on a noncommutative space and $t \longmapsto \exp(t\theta)$ is the one parameter flow of automorphisms (i.e. integral curves) of our noncommutative space generated by θ.*
(ii) Suppose $Der(C)$ be the space of all derivations on C. It is a module over $Z(C)$ such that it has a Lie algebra structure with the Lie bracket given by the commutator with respect to the composition of derivations. [38, 104, 105]

Definition 4.3.9. *Letting $\Omega_{Der}^n(C)$ be the space of all $Z(C)-$multilinear antisymmetric mappings from $Der(C)^n$ into C such that $\Omega_{Der}^0(C) = C$. A differential graded algebra $\Omega_{Der}^\bullet(C) = \bigoplus_{n \geq 0} \Omega_{Der}^n(C)$ can be defined where for each $\omega \in \Omega_{Der}^n(C)$ and $\theta_i \in Der(C)$, its antiderivation differential operator d of degree one is given by*

$$(d\omega)(\theta_0, ..., \theta_n) :=$$

$$\sum_{k=0}^{n}(-1)^k \theta_k \omega(\theta_0, ..., \widehat{\theta_k}, ..., \theta_n) + \sum_{0 \leq r < s \leq n}(-1)^{r+s} \omega([\theta_r, \theta_s], \theta_0, ..., \widehat{\theta_r}, ..., \widehat{\theta_s}, ..., \theta_n).$$

Remark 4.3.10. *(i) $d^2 = 0$.*
(ii) For a given derivation θ of C and $\omega \in \Omega_{Der}^n(C)$, consider an anti-derivation operator i_θ of degree (-1) defined by
$$i_\theta \omega(\theta_1, ..., \theta_{n-1}) = \omega(\theta, \theta_1, ..., \theta_{n-1}).$$
It is observed that
$$i_{\theta_1} i_{\theta_2} + i_{\theta_2} i_{\theta_1} = 0,$$
and if $L_\theta := d \circ i_\theta + i_\theta \circ d$, then
$$L_{\theta_1} i_{\theta_2} - i_{\theta_2} L_{\theta_1} = i_{[\theta_1, \theta_2]},$$
$$L_{\theta_1} L_{\theta_2} - L_{\theta_2} L_{\theta_1} = L_{[\theta_1, \theta_2]}.$$

[38]

Definition 4.3.11. *The noncommutative deRham complex on $Der(C)$ is defined by*

$$DR^\bullet_{Der}(C) := \frac{\Omega^\bullet_{Der}(C)}{[\Omega^\bullet_{Der}(C), \Omega^\bullet_{Der}(C)]}.$$

We know that an algebra C equipped with a bi-derivation Lie bracket $\{.,.\}$ which satisfies in the Jacobi identity determines a *Poisson algebra*. There is a class of derivations on this algebra which contains an essential geometric meaning.

Definition 4.3.12. *For each c in the Poisson algebra C, derivation*

$$ham(c) : x \longmapsto \{c, x\}$$

is called a Hamiltonian derivation (vector field) corresponding to c.

Definition 4.3.13. *For a given Poisson algebra C, a $Z(C)$−bilinear antisymmetric map ω in $\Omega^2_{Der}(C)$ is called non-degenerate, if for any element $c \in C$, there exists a derivation θ_c of C such that for each derivation θ,*

$$\omega(\theta_c, \theta) = \theta(c).$$

In this case derivation $\theta_c = ham(c)$ is unique and the function $\theta \longmapsto i_\theta \omega$ is linear and injective and it is observed that

$$(i_{ham(c)}\omega)(\theta) = \omega(ham(c), \theta) = \theta(c) =: (dc)(\theta)$$

such that $dc : Der(C) \longrightarrow C$ is a 1-form.

Definition 4.3.14. *A closed non-degenerate element ω in $\Omega^2_{Der}(C)$ is called a symplectic structure.*

Applying this symplectic form allows us to define an antisymmetric bilinear bracket on C such that for each $x, y \in C$, it is given by

$$\{x, y\}_\omega := \omega(ham(x), ham(y)). \tag{4.3.3}$$

It satisfies Leibniz law and Jacobi identity and therefore it determines a Poisson bracket on C.

Lemma 4.3.15. *(i) There is a Lie algebra homomorphism from $(C, \{.,.\}_\omega)$ to $(Der(C), [.,.])$.*
(ii) When $Z(C)$−module generated by the set

$$Ham(C) := \{ham(c) : c \in C\}$$

be the entire of the space $Der(C)$, the Jacobi identity for the bracket $\{.,.\}_\omega$ and closed condition for the symplectic structure ω are equivalent. [38, 39, 104, 105]

Proof. One can show that

$$[ham(x), ham(y)] = ham(\{x, y\}_\omega).$$

It means that *ham* plays the role of a Lie homomorphism. □

If a given associative algebra C follows condition (ii) in the lemma 4.3.15, then the Poisson bracket on C is called *non-degenerate* (such that this class of Poisson brackets can provide symplectic structures) and otherwise it is called *degenerate*.

Lemma 4.3.16. *Consider an associative algebra C equipped with a non-degenerate Poisson bracket $\{.,.\}$. There is a symplectic structure ω (on derivations) such that its related Poisson bracket coincides with $\{.,.\}$.* [38, 39]

Proof. Non-degeneracy of the Poisson bracket shows that

$$Z(C).Ham(C) = Der(C).$$

Let θ_1, θ_2 be derivations on C. There exist

$$\{x_1,...,x_m,y_1,...,y_n\} \subset C \quad \{u_1,...,u_m,v_1,...,v_n\} \subset Z(C)$$

such that
$$\theta_1 = \sum_i u_i\, ham(x_i), \quad \theta_2 = \sum_j v_j\, ham(y_j).$$

Define
$$\omega(\theta_1, \theta_2) := \sum_{i,j} u_i v_j \{x_i, y_j\}.$$

$\square$

It is the place to combine the theory of noncommutative differential forms with the Connes-Kreiemr approach to obtain a new family of Hamiltonian systems completely depended upon the renormalization procedure. At first suppose in theory Φ one can perform renormalization with an idempotent Rota-Baxter renormalization map R and the related λ–information $\widetilde{\Phi}_\lambda$ ($\lambda \in \mathbb{K}$). The Lie bracket $[.,.]_\lambda$ determines a Poisson bracket in a natural way where lemma 4.3.16 can induce a symplectic structure ω_λ but for this result we need the non-degeneracy of this Poisson bracket such that in general maybe it does not happen. According to the proof of this lemma and since for the identification of integrals of motion just we should concentrate on Hamiltonian vector fields (derivations), therefore for removal this problem it is enough to work on
- $Z(\widetilde{\Phi}_\lambda)$–module $Der_{Ham}(\widetilde{\Phi}_\lambda)$ generated by the set $Ham(\widetilde{\Phi}_\lambda)$ (i.e. all Hamiltonian derivations of the algebra $\widetilde{\Phi}_\lambda$) instead of the set of all derivations $Der(\widetilde{\Phi}_\lambda)$.
- And restrict the differential graded algebra $\Omega^\bullet_{Der}(\widetilde{\Phi}_\lambda)$ into the differential graded algebra $\Omega^\bullet_{Der_{Ham}}(\widetilde{\Phi}_\lambda)$ such that $\Omega^0_{Der_{Ham}}(\widetilde{\Phi}_\lambda) = \widetilde{\Phi}_\lambda$ and $\Omega^n_{Der_{Ham}}(\widetilde{\Phi}_\lambda)$ is the space of all $Z(\widetilde{\Phi}_\lambda)$–multilinear antisymmetric mappings from $Der_{Ham}(\widetilde{\Phi}_\lambda)^n$ into $\widetilde{\Phi}_\lambda$.

Corollary 4.3.17. *Symplectic forms related to the first class. The differential form*

$$\omega_\lambda : Der_{Ham}(\widetilde{\Phi}_\lambda) \times Der_{Ham}(\widetilde{\Phi}_\lambda) \longrightarrow \widetilde{\Phi}_\lambda$$

in $\Omega^2_{Der_{Ham}}(\widetilde{\Phi}_\lambda)$ given by

$$\omega_\lambda(\theta, \theta') := \sum_{i,j} u_i \circ_\lambda v_j \circ_\lambda [f_i, h_j]_\lambda$$

such that $\{f_1,...,f_m, h_1,...,h_n\} \subset L(H,A)$, $\{u_1,...,u_m,v_1,...,v_n\} \subset Z(\widetilde{\Phi}_\lambda)$ and $\theta = \sum_i u_i \circ_\lambda ham(f_i)$, $\theta' = \sum_j v_j \circ_\lambda ham(h_j)$ is a $Z(\widetilde{\Phi}_\lambda)$–bilinear, antisymmetric, non-degenerate and closed element (i.e. a symplectic structure) .

At second if renormalization map does not have idempotent property or in general, then one can concentrate on the universal Nijenhuis tensor to obtain a Poisson bracket $[.,.]_u$. Maybe it has not non-degeneracy and so for determining a symplectic structure ω_u, it is enough to focus on
- $Z(\widetilde{\Phi}_u)$−module $Der_{Ham}(\widetilde{\Phi}_u)$ generated by the set $Ham(\widetilde{\Phi}_u)$.
- And restrict the differential graded algebra $\Omega^\bullet_{Der}(\widetilde{\Phi}_u)$ into the differential graded algebra $\Omega^\bullet_{Der_{Ham}}(\widetilde{\Phi}_u)$ such that $\Omega^0_{Der_{Ham}}(\widetilde{\Phi}_u) = \widetilde{\Phi}_u$ and $\Omega^n_{Der_{Ham}}(\widetilde{\Phi}_u)$ is the space of all $Z(\widetilde{\Phi}_u)$−multilinear antisymmetric mappings from $Der_{Ham}(\widetilde{\Phi}_u)^n$ into $\widetilde{\Phi}_u$.

Corollary 4.3.18. *Symplectic forms related to the second class.* The differential form

$$\omega_u : Der_{Ham}(\widetilde{\Phi}_u) \times Der_{Ham}(\widetilde{\Phi}_u) \longrightarrow \widetilde{\Phi}_u$$

in $\Omega^2_{Der_{Ham}}(\widetilde{\Phi}_u)$ given by

$$\omega_u(\rho,\mu) := \sum_{i,j} w_i \circ_u z_j \circ_u [g_i, k_j]_u$$

such that $\{g_1, ..., g_m, k_1, ..., k_n\} \subset L(H, \overline{T}(A))$, $\{w_1, ..., w_m, z_1, ..., z_n\} \subset Z(\widetilde{\Phi}_u)$ and $\rho = \sum_i w_i \circ_u ham(g_i)$, $\mu = \sum_j z_j \circ_u ham(k_j)$ is a symplectic structure.

So in short, the Hopf algebraic perturbative renormalization (namely, the couple renormalization map and regularization algebra) determines a family of symplectic structures. Now it is convenient to use the terminology *"symplectic space"* for this new information obtained form the geometric studying of renormalization.

Definition 4.3.19. *Symplectic spaces related to the Hopf algebraic reconstruction of the perturbative renormalization in a renormalizable QFT are introduced by*

$$\Gamma_\lambda := (\widetilde{\Phi}_\lambda, Der_{Ham}(\widetilde{\Phi}_\lambda), \omega_\lambda),$$

and

$$\Gamma_u := (\widetilde{\Phi}_u, Der_{Ham}(\widetilde{\Phi}_u), \omega_u).$$

Lemma 4.3.20. *Suppose θ be a derivation in $Der_{Ham}(\widetilde{\Phi}_\lambda)$ and ρ be a derivation in $Der_{Ham}(\widetilde{\Phi}_u)$. One can show that*

$$L_\theta \omega_\lambda = i_\theta \circ d_\lambda \omega_\lambda + d_\lambda \circ i_\theta \omega_\lambda = d_\lambda(i_\theta \omega_\lambda) = 0$$

and

$$L_\rho \omega_u = i_\rho \circ d_u \omega_u + d_u \circ i_\rho \omega_u = d_u(i_\rho \omega_u) = 0$$

such that d_λ (d_u) is the anti-derivation differential operator for the differential graded algebra $\Omega^\bullet_{Der_{Ham}}(\widetilde{\Phi}_\lambda)$ ($\Omega^\bullet_{Der_{Ham}}(\widetilde{\Phi}_u)$). This derivation is called λ−symplectic (u−symplectic) vector field with respect to the symplectic structure ω_λ (ω_u).

Corollary 4.3.21. *Maps*

$$i_{\omega_\lambda} : Der_{Ham}(\widetilde{\Phi}_\lambda) \longrightarrow DR^1_{Der_{Ham}}(\widetilde{\Phi}_\lambda),$$

$$\theta \longmapsto i_\theta \omega_\lambda$$

and

$$i_{\omega_u} : Der_{Ham}(\widetilde{\Phi}_u) \longrightarrow DR^1_{Der_{Ham}}(\widetilde{\Phi}_u),$$

$$\rho \longmapsto i_\rho \omega_u$$

are well defined, linear and bijection. It means that one can find a bijection between closed one forms in $DR^1_{Der_{Ham}}(\widetilde{\Phi}_\lambda)$ (or $DR^1_{Der_{Ham}}(\widetilde{\Phi}_u)$) and $\lambda-$symplectic (or $u-$symplectic) vector fields.

Proof. For each $\lambda \in \mathbb{K}$, it is enough to know that ω_λ (ω_u) is non-degenerate and closed. $\square$

Definition 4.3.22. *If θ_f^λ be the $\lambda-$symplectic vector field associated with $d_\lambda f$ such that $f \in \Omega^0_{Der_{Ham}}(\widetilde{\Phi}_\lambda) = \widetilde{\Phi}_\lambda$. Then a new Poisson bracket on $\widetilde{\Phi}_\lambda$ is defined by*

$$\{f,g\}_\lambda := i_{\theta_f^\lambda}(d_\lambda g).$$

It is called $\lambda-$Poisson bracket.

Corollary 4.3.23. *(i) For each $\lambda \in \mathbb{K}$, there is a new Lie algebra structure on $\widetilde{\Phi}_\lambda$ such that the map*

$$(\widetilde{\Phi}_\lambda, \{.,.\}_\lambda) \longrightarrow (Der_{Ham}(\widetilde{\Phi}_\lambda), [.,.])$$

is a Lie algebra homomorphism.
(ii) $\lambda-$Poisson bracket $\{.,.\}_\lambda$ is characterized with the symplectic structure ω_λ (and therefore by the Lie brackets $[.,.]_\lambda$). It means that

$$\{f,g\}_\lambda = i_{\theta_f^\lambda}(d_\lambda g) = i_{\theta_f^\lambda} i_{\theta_g^\lambda} \omega_\lambda, \quad i_{[\theta_f^\lambda,\theta_g^\lambda]}\omega_\lambda = d_\lambda i_{\theta_f^\lambda}(d_\lambda g) = d_\lambda \{f,g\}_\lambda.$$

Corollary 4.3.24. *Derivation $\theta^\lambda_{\{f,g\}_\lambda}$ is the unique $\lambda-$symplectic vector field with respect to $d_\lambda\{f,g\}_\lambda$ and it means that $\theta^\lambda_{\{f,g\}_\lambda} = [\theta_f^\lambda, \theta_g^\lambda]$.*

There is a similar process for the symplectic space Γ_u and one can obtain a $u-$Poisson bracket $\{.,.\}_u$ on $\widetilde{\Phi}_u$ induced with the symplectic structure ω_u (and therefore by the Lie bracket $[.,.]_u$).

Remark 4.3.25. *If the Poisson bracket $[.,.]_\lambda$ (or $[.,.]_u$) is non-degenerate, then we can define symplectic structure ω_λ (or ω_u) on the whole space of derivations of $\widetilde{\Phi}_\lambda$ (or $\widetilde{\Phi}_u$) and therefore lemma 4.3.16 shows that the Poisson bracket $\{.,.\}_\lambda$ (or $\{.,.\}_u$) will be coincide with $[.,.]_\lambda$ (or $[.,.]_u$). But in general, might be they are not the same.*

Hamiltonian Systems. In classical mechanics a Hamiltonian system consists of a symplectic manifold (as a configuration space) and a Hamiltonian operator such that a Poisson bracket can be inherited from this information. Then naturally, related motion integrals would be determined while as the result we will enable to identify integrable systems. Here we want to carry out the same procedure to obtain a new family of infinite Hamiltonian systems depended upon renormalizable perturbative theories.

Definition 4.3.26. *(i) For each $\lambda \in \mathbb{K}$ and a fixed $F \in L(H,A)$, the pair (Γ_λ, F) is called $\lambda-$Hamiltonian system.*
(ii) For $G \in L(H, \overline{T}(A))$, the pair (Γ_u, G) is called $u-$Hamiltonian system with respect to the theory Φ and the regularization scheme A.
(iii) F and G are called Hamiltonian functions.

With applying the natural exponential map on the space $Der(\widetilde{\Phi}_\lambda)$ (or $Der(\widetilde{\Phi}_u)$) given by the composition of derivations, we could characterize integrals of motion.

Proposition 4.3.27. *(i) Let Φ be a renormalizable theory with the idempotent renormalization map R and (Γ_λ, F) be a λ-Hamiltonian system with respect to it.*

$$\{f, F\}_\lambda = 0 \iff$$

The map f is constant along the integral curves of θ_F^λ (i.e. 1-parameter flow $\alpha_t^\lambda : t \longmapsto exp(t\theta_F^\lambda)$ of the automorphisms generated by θ_F^λ). The function f is called λ-integral of motion of this system.
(ii) Let Φ be a renormalizable theory with the regularization scheme A and (Γ_u, G) be a u-Hamiltonian system with respect to it.

$$\{g, G\}_u = 0 \iff$$

The map g is constant along the integral curves of ρ_G^u (i.e. 1-parameter flow $\alpha_t^u : t \longmapsto exp(t\rho_G^u)$ of the automorphisms generated by ρ_G^u). The function g is called u-integral of motion for this system.

Proof. With using the *Cartan magic formula* [11], we have

$$\frac{d}{dt}(\alpha_t^\lambda)^*(f) = (\alpha_t^\lambda)^* L_{\theta_F^\lambda} f = (\alpha_t^\lambda)^* i_{\theta_F^\lambda} d_\lambda f$$

$$= (\alpha_t^\lambda)^* i_{\theta_F^\lambda} i_{\theta_f^\lambda} \omega_\lambda = (\alpha_t^\lambda)^* \omega_\lambda(\theta_f^\lambda, \theta_F^\lambda) = (\alpha_t^\lambda)^* \{f, F\}_\lambda = 0.$$

□

And finally, the concept of integrability (based on the behavior of motion integrals) for this class of quantum Hamiltonian systems can be determined in an usual procedure. It means that

Definition 4.3.28. *(i) A λ-Hamiltonian system (Γ_λ, F) is called (n, λ)-integrable, if there exist n linearly independent λ-integrals of motion $f_1 = F, f_2, ..., f_n$ such that $\{f_i, f_j\}_\lambda = 0$.*
(ii) A u-Hamiltonian system (Γ_u, G) is called (n, u)-integrable, if there exist n linearly independent u-integrals of motion $g_1 = G, g_2, ..., g_n$ such that $\{g_i, g_j\}_u = 0$.
(iii) If a λ-Hamiltonian (or u-Hamiltonian) system has infinite linearly independent λ-(or u-)integrals of motion, then it is called infinite dimensional λ-integrable (or infinite dimensional u-integrable) system.

Briefly speaking, we could provide the concept of integrability of Hamiltonian systems in renormalizable QFTs with respect to regularization algebras or idempotent renormalization schemes such that integrals of motion are introduced in consistency of determined Nijenhuis operators. This process enables to reflect the dependency of this style of (integrable) Hamiltonian systems upon the perturbative renormalization. More precisely, there is an interesting chance to check the strong compatibility of this approach with the Connes-Kreimer formalism and it can be done by the renormalization group. Actually, we will show that using Connes-Kreimer renormalization group can guide us to characterize some examples of integrable quantum Hamiltonian systems.

4.4 Integrable systems on the basis of the renormalization group

We saw that how one can reach to a fundamental concept of Hamiltonian formalism from deformation of Connes-Kreimer convolution algebras. It is the place to show further the compatibility of this approach with the Hopf algebraic machinery in terms of the renormalization group. So in this part, we work on the BPHZ prescription to recognize motion integrals depended on characters of the renormalization Hopf algebra and also, we apply the Connes-Kreimer Birkhoff factorization to determine some conditions for components of this factorization on a Feynman rules character ϕ when they are motion integrals of ϕ.

Roughly, here we consider the relation between defined Nijenhuis type motion integrals and physical information underlying minimal subtraction in dimensional regularization. Particularly, we further concentrate on Hamiltonian systems such that their Hamiltonian functions are dimensionally regularized Feynman rules characters or elements of the Connes-Kreimer renormalization group.

Starting with a renormalizable physical theory Φ together with the dimensionally regularized Feynman rules character ϕ underlying the renormalization prescription (A_{dr}, R_{ms}). We know that R_{ms} is a Nijenhuis tensor and so for each $\lambda \in \mathbb{K}$, the map $\Upsilon_{ms,\lambda}$ on $L(H, A_{dr})$ has also this property such that one can extend it to the Lie algebra level. Since we need to study the behavior of the renormalization group, so just we focus on the case $\lambda = 0$ and note that there are similar calculations for other values of λ. Consider the Rota-Baxter map Υ_{ms} where for each $\phi \in L(H, A_{dr})$,

$$\Upsilon_{ms}(\phi) := R_{ms} \circ \phi. \tag{4.4.1}$$

Induce a new associative product and a Lie bracket on $L(H, A_{dr})$ given by

$$\phi_1 \circ_0 \phi_2 := R_{ms} \circ \phi_1 * \phi_2 + \phi_1 * R_{ms} \circ \phi_2 - R_{ms} \circ (\phi_1 * \phi_2). \tag{4.4.2}$$

$$[\phi_1, \phi_2]_0 := [R_{ms} \circ \phi_1, \phi_2] + [\phi_1, R_{ms} \circ \phi_2] - R_{ms} \circ ([\phi_1, \phi_2]). \tag{4.4.3}$$

Theorem 4.2.1 shows that

$$R_{ms} \circ [\phi_1, \phi_2]_0 = [R_{ms} \circ \phi_1, R_{ms} \circ \phi_2]. \tag{4.4.4}$$

Definition 4.4.1. *The* $0-$*information based on the theory* Φ *and with respect to the map* Υ_{ms} *is defined by*

$$(\widetilde{\Phi}_0, [.,.]_0) := (L(H, A_{dr}), \circ_0, \Upsilon_{ms}, [.,.]_0).$$

Corollary 4.3.17 determines a symplectic structure for this $0-$information. For arbitrary derivations $\theta = \sum_i u_i \circ_0 ham(f_i)$ and $\theta' = \sum_j v_j \circ_0 ham(h_j)$ in $Der_{Ham}(\widetilde{\Phi}_0)$ where $f_1, ..., f_m, h_1, ..., h_n$ are in $L(H, A_{dr})$ and $u_1, ..., u_m, v_1, ..., v_n$ are in $Z(\widetilde{\Phi}_0)$, we have

$$\omega_0(\theta, \theta') := \sum_{i,j} u_i \circ_0 v_j \circ_0 [f_i, h_j]_0. \tag{4.4.5}$$

Definition 4.4.2. *The symplectic space related to the theory* Φ *and the renormalization map* R_{ms} *is given by*

$$\Gamma_0 := (\widetilde{\Phi}_0, Der_{Ham}(\widetilde{\Phi}_0), \omega_0).$$

By choosing a Hamiltonian function $F \in L(H, A_{dr})$, one can have a $0-$Hamiltonian system (Γ_0, F) with respect to the theory and from proposition 4.3.27, a $0-$integral of motion for this system is an element $f \in L(H, A_{dr})$ such that

$$\{f, F\}_0 = 0. \tag{4.4.6}$$

Let θ_F^0, θ_f^0 be the 0−symplectic vector fields with respect to d_0F, d_0f. Therefore

$$i_{\theta_F^0}\omega_0 = d_0F, \quad i_{\theta_f^0}\omega_0 = d_0f, \qquad (4.4.7)$$

$$\{f, F\}_0 = i_{\theta_F^0} i_{\theta_f^0} \omega_0 = \omega_0(\theta_f^0, \theta_F^0) = [f, F]_0 = 0. \qquad (4.4.8)$$

Lemma 4.4.3. *Let $f \in L(H, A_{dr})$ be a 0−integral of motion for the system (Γ_0, F). Then we have*

$$R_{ms} \circ f * R_{ms} \circ F = R_{ms} \circ F * R_{ms} \circ f.$$

Proof. By the formula (4.4.4),

$$R_{ms} \circ [f, F]_0 = [R_{ms} \circ f, R_{ms} \circ F] \implies$$

$$[R_{ms} \circ f, R_{ms} \circ F] = R_{ms}(0) = 0 \implies$$

$$R_{ms} \circ f * R_{ms} \circ F - R_{ms} \circ F * R_{ms} \circ f = 0.$$

□

Lemma 4.4.4. *The equation*

$$R_{ms} \circ f * F - F * R_{ms} \circ f + f * R_{ms} \circ F - R_{ms} \circ F * f - R_{ms} \circ (f * F) + R_{ms} \circ (F * f) = 0$$

determines a necessary and sufficient condition to characterize 0−integrals of motion for the 0−Hamiltonian system (Γ_0, F) with respect to the minimal subtraction scheme in dimensional regularization.

Proof. It is proved by the equation (4.4.3). Because it shows that for a 0−integral of motion f we have

$$[f, F]_0 = 0 \iff$$

$$[R_{ms} \circ f, F] + [f, R_{ms} \circ F] - R_{ms} \circ [f, F] = 0.$$

□

If the Hamiltonian function of this system is the dimensionally regularized Feynman rules character ϕ of the theory, then one can obtain interesting relations between the components of the Birkhoff factorization of ϕ and 0−integrals of motion of this system.

Corollary 4.4.5. *For the dimensionally regularized Feynman rules character $\phi \in L(H, A_{dr})$, let f be a 0−integral of motion for the 0−Hamiltonian system (Γ_0, ϕ). Then for each Feynman diagram $\Gamma \in \ker \epsilon_H$, f satisfies in the equation*

$$\sum_\gamma R_{ms}(f(\gamma)) R_{ms}(\phi(\frac{\Gamma}{\gamma})) = \sum_\gamma R_{ms}(\phi(\gamma)) R_{ms}(f(\frac{\Gamma}{\gamma})).$$

Proof. For each $\Gamma \in \ker \epsilon_H$, its coproduct is given by

$$\Delta(\Gamma) = \Gamma \otimes 1 + 1 \otimes \Gamma + \sum_{\gamma \subset \Gamma} \gamma \otimes \frac{\Gamma}{\gamma}$$

58

such that the sum is over all disjoint unions of primitive 1PI superficially divergent proper subgraphs. Let f be a 0−integral of motion of this system. Renormalization coproduct on Feynman diagrams shows us that

$$(R_{ms} \circ f * R_{ms} \circ \phi)(\Gamma) =$$
$$R_{ms}(f(1))R_{ms}(\phi(\Gamma)) + R_{ms}(f(\Gamma))R_{ms}(\phi(1)) + \sum_{\gamma} R_{ms}(f(\gamma))R_{ms}(\phi(\frac{\Gamma}{\gamma})),$$
$$(R_{ms} \circ \phi * R_{ms} \circ f)(\Gamma) =$$
$$R_{ms}(\phi(\Gamma))R_{ms}(f(1)) + R_{ms}(\phi(1))R_{ms}(f(\Gamma)) + \sum_{\gamma} R_{ms}(\phi(\gamma))R_{ms}(f(\frac{\Gamma}{\gamma})).$$

Therefore with help of the corollary 4.4.3, the mentioned formula should be obtained such that the sum has finite terms where each term contains pole parts of Laurent series. □

Let f be a 0−integral of motion for the system (Γ_0, ϕ) such that ϕ is the Feynman rules character of the theory. Theorem 3.2.6 shows that for each primitive 1PI (superficially divergent) proper subgraph γ of the Feynman diagram Γ,

$$\phi_-(\gamma) = -R_{ms}(\phi(\gamma)). \qquad (4.4.9)$$

Therefore with the help of corollary 4.4.5 and (4.4.9), the relations between components of the factorization of ϕ and this 0−integral of motion can be available. For each Feynman diagram Γ which is a disjoint union of primitive 1PI diagrams, we have

$$\sum_{\gamma} R_{ms}(f(\gamma))R_{ms}(\phi(\frac{\Gamma}{\gamma})) = \sum_{\gamma} -\phi_-(\gamma)R_{ms}(f(\frac{\Gamma}{\gamma}))$$
$$\Longrightarrow$$
$$\sum_{\gamma} R_{ms}(f(\gamma))R_{ms}(\phi(\frac{\Gamma}{\gamma}))R_{ms}(f(\frac{\Gamma}{\gamma}))^{-1} + \phi_-(\gamma) = 0$$
$$\Longrightarrow$$

$$\sum_{\gamma} \phi_-(\gamma) = -\sum_{\gamma} R_{ms}(f(\gamma))R_{ms}(\phi(\frac{\Gamma}{\gamma}))R_{ms}(f(\frac{\Gamma}{\gamma}))^{-1}, \qquad (4.4.10)$$

$$\sum_{\gamma} \phi_+(\gamma) = \sum_{\gamma} \phi(\gamma) - \sum_{\gamma} R_{ms}(f(\gamma))R_{ms}(\phi(\frac{\Gamma}{\gamma}))R_{ms}(f(\frac{\Gamma}{\gamma}))^{-1}. \qquad (4.4.11)$$

Corollary 4.4.6. *(i) Applying (4.4.10), (4.4.11) and equations mentioned in the theorem 3.2.6, new representations from the factorization components of a Feynman rules character ϕ (based on the 0−integral of motion f) can be determined.*

(ii) Putting these equations in the lemma 3.2.8 and then using corollary 4.4.5, new relations between this 0−integral of motion and renormalization group and also β−function of the theory can be investigated.

Now we have a chance to search more geometrical meanings in the Feynman rules character ϕ and it can be performed based on motion integral condition for components of Birkhoff factorization of this character. It is necessary to emphasize that since the renormalization method is fixed and the factorization is unique, therefore consideration of this possibility helps us to familiar with some more hidden geometrical structures in a quantum field theory which provide the required integral conditions.

Corollary 4.4.7. *Suppose the negative part ϕ_- of the Birkhoff decomposition of the dimensionally regularized Feynman rules character $\phi \in L(H, A_{dr})$ of a given theory Φ be a 0−integral of motion for the 0−Hamiltonian system (Γ_0, ϕ). Then for each Feynman diagram Γ which is a disjoint union of primitive 1PI diagrams, we have*

$$R_{ms}(\sum_{\gamma} \sum_{\gamma' \subset \Gamma_\gamma} \phi_-(\gamma')\phi(\frac{\Gamma_\gamma}{\gamma'})) = 0.$$

Proof. If ϕ_- be a 0−integral of motion, then by given conditions in corollary 4.4.5 for the Feynman diagram Γ one should have

$$\sum_{\gamma} R_{ms}(\phi_-(\gamma))R_{ms}(\phi(\frac{\Gamma}{\gamma})) = \sum_{\gamma} R_{ms}(\phi(\gamma))R_{ms}(\phi_-(\frac{\Gamma}{\gamma})).$$

By (4.4.9) and since R_{ms} is an idempotent linear map, it can be seen that

$$\sum_{\gamma} -R_{ms}(R_{ms}(\phi(\gamma)))R_{ms}(\phi(\frac{\Gamma}{\gamma})) =$$

$$\sum_{\gamma} -R_{ms}(\phi(\gamma))R_{ms}(\phi(\frac{\Gamma}{\gamma})) = \sum_{\gamma} R_{ms}(\phi(\gamma))R_{ms}(\phi_-(\frac{\Gamma}{\gamma}))$$

$$\Longrightarrow \sum_{\gamma} -R_{ms}(\phi(\frac{\Gamma}{\gamma})) = \sum_{\gamma} R_{ms}(\phi_-(\frac{\Gamma}{\gamma})) = \sum_{\gamma} \phi_-(\frac{\Gamma}{\gamma}).$$

Set $\Gamma_\gamma := \frac{\Gamma}{\gamma}$. By theorem 3.2.6,

$$\sum_{\gamma} \phi_-(\Gamma_\gamma) = -R_{ms} \sum_{\gamma} (\phi(\Gamma_\gamma) + \sum_{\gamma' \subset \Gamma_\gamma} \phi_-(\gamma')\phi(\frac{\Gamma_\gamma}{\gamma'}))$$

$$\Longrightarrow \sum_{\gamma} R_{ms}(\phi(\Gamma_\gamma) + \sum_{\gamma' \subset \Gamma_\gamma} \phi_-(\gamma')\phi(\frac{\Gamma_\gamma}{\gamma'})) = \sum_{\gamma} R_{ms}(\phi(\Gamma_\gamma))$$

such that the sum is over all unions of primitive 1PI proper superficially divergent subgraphs of Γ_γ. □

Using theorems 3.2.5, 3.2.6 and the corollary 4.4.4 and since R_{ms} is an idempotent map lead us to produce more explicit conditions for the components of decomposition of ϕ which play the role of 0−integrals of motion for the system (Γ_0, ϕ).

Corollary 4.4.8. *For the 0−Hamiltonian system (Γ_0, ϕ) such that ϕ is the dimensionally regularized Feynman rules character (in $L(H, A_{dr})$) of the theory Φ, the components of the Birkhoff factorization of ϕ are 0−integrals of motion for the system if and only if*

(i) ϕ_- satisfies in the equation

$$\phi_+ - \phi * \phi_- + \phi_- * R_{ms} \circ \phi - R_{ms} \circ \phi * \phi_- + R_{ms} \circ (\phi * \phi_-) = 0.$$

(ii) ϕ_+ satisfies in the equation

$$\phi_+ * R_{ms} \circ \phi - R_{ms} \circ \phi * \phi_+ - R_{ms} \circ (\phi_+ * \phi) + R_{ms} \circ (\phi * \phi_+) = 0.$$

Proof. For the negative part ϕ_- we have

$$R_{ms} \circ \phi_- * \phi - \phi * R_{ms} \circ \phi_- + \phi_- * R_{ms} \circ \phi - R_{ms} \circ \phi * \phi_- - R_{ms} \circ (\phi_- * \phi) + R_{ms} \circ (\phi * \phi_-) = 0$$

$$\Longleftrightarrow$$

$$\phi_- * \phi - \phi * \phi_- + \phi_- * R_{ms} \circ \phi - R_{ms} \circ \phi * \phi_- - R_{ms} \circ (\phi_- * \phi) + R_{ms} \circ (\phi * \phi_-) = 0, \quad \phi = \phi_-^{-1} * \phi_+$$

$$\Longleftrightarrow$$

$$\phi_+ - \phi * \phi_- + \phi_- * R_{ms} \circ \phi - R_{ms} \circ \phi * \phi_- - R_{ms} \circ \phi_+ + R_{ms} \circ (\phi * \phi_-) = 0.$$

For the positive part ϕ_+ we have

$$R_{ms} \circ \phi_+ * \phi - \phi * R_{ms} \circ \phi_+ + \phi_+ * R_{ms} \circ \phi - R_{ms} \circ \phi * \phi_+ - R_{ms} \circ (\phi_+ * \phi) + R_{ms} \circ (\phi * \phi_+) = 0.$$

Now since $R_{ms} \circ \phi_+ = 0$, the proof is completed. □

It is good place to consider the behavior of the renormalization group with respect to the 0−Hamiltonian system (Γ_0, ϕ). With help of this group (associated to the Feynman rules character ϕ), we are going to introduce an infinite dimensional 0−integrable system with respect to the theory Φ.

Lemma 4.4.9. *For each t, an element F_t of the renormalization group is a 0−integral of motion for the 0−Hamiltonian system (Γ_0, ϕ) if and only if it satisfies in*

$$F_t * R_{ms} \circ \phi - R_{ms} \circ \phi * F_t - R_{ms} \circ (F_t * \phi) + R_{ms} \circ (\phi * F_t) = 0.$$

Proof. Since for each t and Feynman diagram Γ, $F_t(\Gamma)$ is a polynomial in t, therefore $R_{ms}(F_t(\Gamma)) = 0$. With notice to (4.4.4), we have

$$R_{ms} \circ F_t * \phi - \phi * R_{ms} \circ F_t + F_t * R_{ms} \circ \phi - R_{ms} \circ \phi * F_t - R_{ms} \circ (F_t * \phi) + R_{ms} \circ (\phi * F_t) = 0.$$

□

For arbitrary elements F_t, F_s of the renormalization group of each given renormalizable theory Φ, one can observe that

$$\{F_t, F_s\}_0 = [R_{ms} \circ F_t, F_s] + [F_t, R_{ms} \circ F_s] - R_{ms} \circ ([F_t, F_s]). \quad (4.4.12)$$

Since the renormalization group is a 1-parameter subgroup of $G(\mathbb{C})$ (i.e. $F_t * F_s = F_{t+s}$), therefore it is easy to see that

$$\{F_t, F_s\}_0 = 0. \quad (4.4.13)$$

It shows that the renormalization group can give us an integrable system and this fact turns out a strong relation between (the given Nijenhuis type) integrals of motion and the Connes-Kreimer renormalization group.

Corollary 4.4.10. *For the renormalizable physical theory Φ, let $\phi \in G(A_{dr})$ be its dimensionally regularized Feynman rules character and $\{F_t\}_t$ be the renormalization group with respect to this character. For each arbitrary element F_{t_0} of the renormalization group, 0−Hamiltonain system (Γ_0, F_{t_0}) is an infinite dimensional 0−integrable system.*

Proof. The renormalization group contains infinite linearly independent 0−integrals of motion F_t for the system (Γ_0, F_{t_0}). □

4.5 Baditoiu-Rosenberg framework: The continuation of the standard process

In [10] the authors (by working on the Lie group of diffeographisms and its related Lie algebra of infinitesimal characters) improve the study of infinite dimensional Lie algebras and factorization problem to the level of the Connes-Kreimer theory. In this part, we focus on their strategy and apply their results to consider factorization problem on the previously introduced noncommutative algebras (deformed by Nijenhuis maps) such that consequently, a new family of integrals of motion will be determined in a natural manner [98].

The Lie brackets $[.,.]_\lambda$ make available another procedure to study integrable systems at this level namely, identifying equations of motion and integral curves from a Lax pair equation. Set $C_\lambda^{dr} := (L(H, A_{dr}), \circ_\lambda)$ and supposing $C_\lambda = C_\lambda^{dr} \oplus C_\lambda^{dr*}$ be its related semisimple trivial Lie bialgebra such that $C_\lambda^{dr} := (C_\lambda^{dr}, [.,.]_\lambda)$.

Corollary 4.5.1. C_λ is the associated Lie algebra of the Lie group $\widetilde{C}_\lambda := C_\lambda^{dr} \rtimes_\sigma C_\lambda^{dr*}$ such that

$$\sigma : C_\lambda^{dr} \times C_\lambda^{dr*} \longrightarrow C_\lambda^{dr*}, \quad (f, X) \longmapsto Ad^*(f)(X).$$

Proof. It is directly obtained from [10]. □

Definition 4.5.2. *The loop algebra of C_λ is defined by the set*

$$LC_\lambda := \{F(c) = \sum_{j=-\infty}^{\infty} c^j F_j, F_j \in C_\lambda\}$$

such that naturally, one can define the Lie bracket

$$[\sum c^i F_i, \sum c^j G_j] := \sum_k c^k \sum_{i+j=k} [F_i, G_j]_\lambda$$

on it.

Decompose this set of formal power series into two parts

$$LC_{\lambda,+} = \{\sum_{j=0}^{\infty} c^j F_j\}, \quad LC_{\lambda,-} = \{\sum_{j=-\infty}^{-1} c^j F_j\} \quad (4.5.1)$$

and let $P_\pm$ are the natural projections on these components where $P := P_+ - P_-$.

Corollary 4.5.3. *For a given Casimir function v on LC_λ, integral curve $\Lambda(t)$ of the Lax pair equation $\frac{d\Lambda}{dt} = [M, \Lambda]$ where for $F(\lambda) = F(0)(\lambda) \in LC_\lambda$, $M = \frac{1}{2}P(I(dv(F(c)))) \in LC_\lambda$ is given by*

$$\Lambda(t) = Ad^*_{\widetilde{LC_\lambda}} \gamma_\pm(t).\Lambda(0)$$

such that the smooth curves $\gamma_\pm$ are the answers of the Birkhoff factorization

$$exp(-tX) = \gamma_-^{-1}(t)\gamma_+(t)$$

where $X = I(dv(F(c))) \in LC_\lambda$.

Proof. It is easily calculated from [10, 95]. □

Remark 4.5.4. *(i) It is important to know that one can project the above Lax pair equation to an equation on loop algebra of the original Lie algebra $\mathcal{C}_\lambda^{dr}$.*
(ii) Based on the induced Nijenhuis type symplectic structures, one can introduce a symplectic space on the loop algebra $L\mathcal{C}_\lambda$. Next, depended motion integrals and therefore integrable Hamiltonian systems on this loop algebra can be determined.

4.6 Fixed point equations

We observed that the Connes-Kreimer recalculation of the BPHZ renormalization depends strongly on components of the Birkhoff factorization of dimensionally regularized Feynman rules characters and on the other hand, the existence of this factorization is supported originally by Atkinson's theorem [40].

Theorem 4.6.1. *Let A be an associative algebra over the field $\mathbb{K}$ and $R : A \longrightarrow A$ be a linear map. The pair (A, R) is a Rota-Baxter algebra if and only if the algebra A has a Birkhoff factorization. It means that there is a Cartesian product*

$$(R(A), \widetilde{R}(A)) \subset A \times A$$

such that
- *It is a subalgebra of $A \times A$,*
- *Each $x \in A$ admits a unique decomposition $x = R(x) \oplus \widetilde{R}(x)$. [2, 30]*

This theorem can determine explicit inductive formulaes for these components of the Birkhoff factorization of characters on the Connes-Kreimer Hopf algebra of rooted trees such that at this level, the notion of a decomposition for Lie algebras will be available. With attention to the BPHZ prescription, in this part we are going to apply these recursive equations to introduce a new family of fixed point equations related to the motion integral condition on components of a Feynman rules character. We will show that this type of equations are formalized by Bogoliubov character and BCH series. Furthermore, the behavior of β-function and renormalization group underlying these equations will be considered [98]. It should be important to note that because of the universality of the Connes-Kreimer Hopf algebra (with respect to the Hochschild cohomology theory), the study of this family of fixed point equations at this level can be lifted to other Hopf algebras of renormalizable theories.

Lemma 4.6.1. *Consider the group $\mathrm{char}_{A_{dr}} H_x$ ($x = lrt, rt$) of characters on ladder trees (rooted trees) with its related Lie algebra $\partial\, \mathrm{char}_{A_{dr}} H_x$. This Lie algebra is generated by derivations Z^t indexed by ladder tree (rooted tree) t and defined by the natural paring*

$$< Z^t, s > = \delta_{t,s}.$$

Based on the bijection between $\mathrm{char}_{A_{dr}} H_x$ and $\partial\, \mathrm{char}_{A_{dr}} H_x$, for each character g with the corresponding derivation Z_g, we have

$$g = exp^*(Z_g) = \sum_{n=0}^{\infty} \frac{Z_g^{*n}}{n!}. \qquad (4.6.1)$$

Lemma 4.6.2. *(i) The Lie algebra $\partial\, char_{A_{dr}}\, H_x$ together with the map $\mathcal{R}: f \longmapsto R_{ms} \circ f$ define a Lie Rota-Baxter algebra.*
(ii) Each derivation Z of this Lie algebra has a unique decomposition

$$Z = \mathcal{R}(Z) \oplus \widetilde{\mathcal{R}}(Z).$$

(iii) Idempotent property of R provides a decomposition $A_{dr} = A_{dr}^+ \oplus A_{dr}^-$ such that it can be extended to $\partial\, char_{A_{dr}}\, H_x$ and it means that

$$\partial\, char_{A_{dr}}\, H_x = (\partial\, char_{A_{dr}}\, H_x)_+ \oplus (\partial\, char_{A_{dr}}\, H_x)_-$$

where

$$(\partial\, char_{A_{dr}}\, H_x)_+ := \widetilde{\mathcal{R}}(\partial\, char_{A_{dr}}\, H_x), \quad (\partial\, char_{A_{dr}}\, H_x)_- := \mathcal{R}(\partial\, char_{A_{dr}}\, H_x)$$

[27, 28, 29].

For a fixed character $g \in C_\lambda^x$, let f be its integral of motion. Product $\circ_\lambda$, its related λ–Poisson bracket and motion integral condition show that

$$\{f, g\}_\lambda = i_{\theta_g^\lambda} i_{\theta_f^\lambda} \omega_\lambda = \omega_\lambda(\theta_f^\lambda, \theta_g^\lambda) = [f, g]_\lambda = 0. \qquad (4.6.2)$$

And so it is apparently observed that

$$[\mathcal{R}_\lambda(f), g] + [f, \mathcal{R}_\lambda(g)] - \mathcal{R}_\lambda([f, g]) = 0 \iff$$

$$\mathcal{R}_\lambda(f) * g - g * \mathcal{R}_\lambda(f) + f * \mathcal{R}_\lambda(g) - \mathcal{R}_\lambda(g) * f - \mathcal{R}_\lambda(f * g) + \mathcal{R}_\lambda(g * f) = 0. \qquad (4.6.3)$$

Proposition 4.6.3. *For the given character $g \in C_\lambda^x$ with the Birkhoff factorization (g_-, g_+),*
(i) $g = g_-^{-1} * g_+$, $R_{ms} \circ g_- = g_-$, $R_{ms} \circ g_+ = 0$.
(ii) The negative component g_- is an integral of motion for g iff

$$g_+ - g * g_- + (1+\lambda)g_- * \mathcal{R}(g) - (1+\lambda)\mathcal{R}(g) * g_- + (1+\lambda)\mathcal{R}(g * g_-) = 0.$$

(iii) The positive component g_+ is an integral of motion for g iff

$$-\lambda g_+ * g + \lambda g * g_+ + (1+\lambda)g_+ * \mathcal{R}(g) - (1+\lambda)\mathcal{R}(g) * g_+ - (1+\lambda)\mathcal{R}(g_+ * g) + (1+\lambda)\mathcal{R}(g * g_+) = 0.$$

Atkinson theorem determines very interesting recursive representation from components of decomposition of characters on rooted trees such that because of its practical structure in calculating physical information, this representation is called *tree renormalization*.

Theorem 4.6.2. *(i) **Ladder tree renormalization**. Each arbitrary character $\phi \in char_{A_{dr}}\, H_{lrt}$ has a unique Birkhoff factorization (ϕ_-^{-1}, ϕ_+) such that*

$$\phi = exp^*(\mathcal{R}(Z_\phi) + \widetilde{\mathcal{R}}(Z_\phi)) = exp^*(\mathcal{R}(Z_\phi)) * exp^*(\widetilde{\mathcal{R}}(Z_\phi)),$$

$$\phi_- = exp^*(-\mathcal{R}(Z_\phi)), \quad \phi_+ = exp^*(\widetilde{\mathcal{R}}(Z_\phi)).$$

(ii) **Rooted tree renormalization.** Each arbitrary character $\psi \in char_{A_{dr}} H_{rt}$ has a unique Birkhoff factorization (ψ_-^{-1}, ψ) such that

$$\psi = exp^*(Z_\psi) = exp^*(\mathcal{R}(\chi(Z_\psi))) * exp^*(\widetilde{\mathcal{R}}(\chi(Z_\psi))),$$

$$\psi_- = exp^*(-\mathcal{R}(\chi(Z_\psi))), \quad \psi_+ = exp^*(\widetilde{\mathcal{R}}(\chi(Z_\psi)))$$

where the infinitesimal character χ is characterized with the BCH series. [27, 28, 29, 30]

Now with applying representations given in the theorem 4.6.2 and conditions given in the proposition 4.6.3, one can obtain new equations at the level of Lie algebra for while these components are motion integrals of a given character in the algebra C_χ^x.

In addition, there is another procedure to deform the initial product (based on double Rota-Baxter structures) such that this new deformed product can induce another representation from components.

Definition 4.6.4. *Rota-Baxter map $\mathcal{R}$ deforms the convolution product $*$ to define a well known associative product on the set $L(H_x, A_{dr})$ given by*

$$f *_\mathcal{R} g := f * \mathcal{R}(g) + \mathcal{R}(f) * g - f * g.$$

Lemma 4.6.5. *(i) Information $C_\mathcal{R}^x := (L(H_x, A_{dr}), *_\mathcal{R}, \mathcal{R})$ is a Rota-Baxter algebra with the corresponding $\mathcal{R}-$bracket*

$$[f, g]_\mathcal{R} = [f, \mathcal{R}(g)] + [\mathcal{R}(f), g] - [f, g].$$

(ii) For each infinitesimal character Z, it can be seen that

$$exp^*(\mathcal{R}(Z)) = \mathcal{R}(exp^{*\mathcal{R}}(Z)), \quad exp^*(\widetilde{\mathcal{R}}(Z)) = -\widetilde{\mathcal{R}}(exp^{*\mathcal{R}}(-Z)).$$

Corollary 4.6.6. *For the given character $g \in C_\mathcal{R}^x$ with the Birkhoff factorization (g_-, g_+), the components are integrals of motion for g iff*
(i) $[g_-, \mathcal{R}(g)] = 0$,
(ii) $[g_+, \mathcal{R}(g)] - [g_+, g] = 0$,
respectively.

Proof. It is proved on the basis of the definition 4.6.4 and applying proposition 4.4.8 and lemma 4.6.5. □

Corollary 4.6.7. *Lie algebra version of the given equations in the corollary 4.6.6 are reformulated by*

$$(i) \; \mathcal{R}(exp^{*\mathcal{R}}(-\chi(Z_\psi))) * \mathcal{R}(exp^*(Z_\psi)) - \mathcal{R}(exp^*(Z_\psi)) * \mathcal{R}(exp^{*\mathcal{R}}(-\chi(Z_\psi))) = 0,$$

$$(ii) - \widetilde{\mathcal{R}}(exp^{*\mathcal{R}}(-\chi(Z_\psi))) * \mathcal{R}(exp^*(Z_\psi)) + \mathcal{R}(exp^*(Z_\psi)) * \widetilde{\mathcal{R}}(exp^{*\mathcal{R}}(-\chi(Z_\psi)))$$

$$+ \widetilde{\mathcal{R}}(exp^{*\mathcal{R}}(-\chi(Z_\psi))) * exp^*(Z_\psi) - exp^*(Z_\psi) * \widetilde{\mathcal{R}}(exp^{*\mathcal{R}}(-\chi(Z_\psi))) = 0.$$

Now we want to relate a family of fixed point equations to the equations induced in the corollary 4.6.7 and because of that we need a new operator.

Definition 4.6.8. *The morphism*

$$b[\psi] := exp^{*\mathcal{R}}(-\chi(Z_\psi))$$

is called Bogoliubov character.

Lemma 4.6.9. *One can approximate Bogoliubov character with the formula*

$$\mathcal{R}(b[\psi]) = -R_{ms} \circ \{exp^*(Z_\psi) + \alpha_\psi\}$$

such that

$$\alpha_\psi := \sum_{n \geq 0} \frac{1}{n!} \sum_{j=1}^{n-1} \frac{n!}{j!(n-j)!} \mathcal{R}(-\chi(Z_\psi))^{*(n-j)} * Z_\psi^{*j}.$$

[29, 30]

Corollary 4.6.10. *One can rewrite equations in the corollary 4.6.7 based on the given estimation in the lemma 4.6.9. We have*

$$(i)' \quad -\mathcal{R}(\psi + \alpha_\psi) * \mathcal{R}(exp^*(Z_\psi)) + \mathcal{R}(exp^*(Z_\psi)) * \mathcal{R}(\psi + \alpha_\psi) = 0$$

and

$$(ii)' \quad \widetilde{\mathcal{R}}(\psi + \alpha_\psi) * \mathcal{R}(exp^*(Z_\psi)) - \mathcal{R}(exp^*(Z_\psi)) * \widetilde{\mathcal{R}}(\psi + \alpha_\psi) - \widetilde{\mathcal{R}}(\psi + \alpha_\psi) * exp^*(Z_\psi)$$

$$+ exp^*(Z_\psi) * \widetilde{\mathcal{R}}(\psi + \alpha_\psi) = 0.$$

Furthermore, we saw that Birkhoff factorization of characters of the Connes-Kreimer Hopf algebra are identified with the special infinitesimal character χ such that for each infinitesimal character $Z \in \partial char_{A_{dr}} H_{rt}$, it is given by

$$\chi(Z) = Z + \sum_{k=1}^{\infty} \chi_Z^{(k)}. \qquad (4.6.4)$$

The sum is a finite linear combination of infinitesimal characters and $\chi_Z^{(k)}$'s are determined by solution of the fixed point equation

$$E: \quad \chi(Z) = Z - \sum_{k=1}^{\infty} c_k K^{(k)}(\mathcal{R}(\chi(Z)), \widetilde{\mathcal{R}}(\chi(Z))) \qquad (4.6.5)$$

such that terms $K^{(k)}$s are calculated with the BCH series [26, 27, 28, 29]. With putting the equation E in the Bogoliubov character and then applying the result 4.6.10, one can reformulate the motion integral condition for components of a given character with respect to the fixed point equation E.

Proposition 4.6.11. *For a given Feynman rules character $\psi \in C_{\mathcal{R}}^{rt}$, if each of its Birkhoff factorization's components is an integral of motion for ψ, then a class of fixed point equations will be determined.*

Probably discussion about renormalization group and its infinitesimal generator (i.e. beta function) can be interested at this level. We saw that these physical information are defined based on the grading operator Y (that providing the scaling evolution of the coupling constant) such that this element exists from the extension of the Lie algebra $\partial\ char_{A_{dr}}H_{rt}$ by an element Z^0 where for each rooted tree t, we have

$$[Z^0, Z^t] = Y(Z^t) = |t|Z^t. \tag{4.6.6}$$

Lemma 4.6.12. *For each character $\psi \in char_{A_{dr}}H_{rt}$, its related β-function is given by*

$$\beta(\psi) = \psi_- * [Z^0, \psi_-^{-1}] = \psi_- * Z^0 * \psi_-^{-1} - Z^0.$$

[29]

It is obvious that with applying the exponential map, its related renormalization group is determined by

$$F_t = exp^*(t\beta). \tag{4.6.7}$$

On the other hand, we know that for each $t \in \mathbb{R}$, F_t is a character given by a polynomial of the variable t and therefore

$$R_{ms} \circ F_t = 0 \tag{4.6.8}$$

([22, 28, 93]). Proposition 4.4.9 shows that

Corollary 4.6.13. *Each element of the renormalization group plays the role of an integral of motion for ψ in the algebras C_λ^x and $C_\mathcal{R}^x$ if and only if*

$$-\lambda[F_t, \psi] + [F_t, \mathcal{R}_\lambda(\psi)] - \mathcal{R}_\lambda([\psi, F_t]) = 0,$$

$$[F_t, \mathcal{R}(\psi)] - [F_t, \psi] = 0,$$

respectively.

The second condition in corollary 4.6.13 introduces a fixed point equation depended on the Feynman rules character ψ and its related β-function.

Corollary 4.6.14. *For a given Feynman rules character $\psi \in C_\mathcal{R}^{rt}$, an element F_t of the related renormalization group plays the role of integral of motion for ψ iff the β-function satisfies in the fixed point equation*

$$[exp^*(t\beta), \mathcal{R}\{exp^*(\mathcal{R}(E)) * exp^*(\widetilde{\mathcal{R}}(E))\}] - [exp^*(t\beta), exp^*(\mathcal{R}(E)) * exp^*(\widetilde{\mathcal{R}}(E))] = 0.$$

And finally, it should be remarked that because of the one parameter property of the renormalization group (i.e. $F_t * F_s = F_{t+s}$), one can easily show that for a fixed t_0 in the cases C_0^x and $C_\mathcal{R}^x$, each F_t is an integral of motion for F_{t_0}. So it is reasonable to expect infinite integrable systems related with algebra $C_\mathcal{R}^x$.

Chapter 5

Connes-Marcolli Theory

The Connes-Kreimer conceptional interpretation of renormalization theory could provide a new Hopf algebraic reconstruction from physical information. In continuation of this approach, Connes and Marcolli initiated a new categorical framework based on geometric objects to describe renormalizable physical theories and in fact, they showed that there is a fundamental mathematical construction hidden inside of divergences. According to their programme, the BPHZ renormalization (i.e. minimal subtraction scheme in dimensional regularization) determines a principal bundle such that solutions of classes of differential equations related to some particular flat connections on this bundle can encode counterterms. Then with collecting all of these connections into a category, they introduced a new categorical formalism to consider physical theories such that it leaded to a universal treatment with respect to the Connes-Kreimer theory. [18, 19, 20, 21, 22]

Universal singular frame is evidently one of the most important foundations of this machinery where it contains a deep physical meaning in the sense that with working at this level, all of the divergences can be disappeared and so one can provide a finite theory. Moreover, one can find significance relations between this special frame and noncommutative geometry based on the local index theorem. [18, 21]

Nowadays a new theory of mixed Tate motives is developed on the basis of the Connes-Marcolli formalism which it reports about very desirable connections between theory of motives and quantum field theory. [18]

In this chapter, we want to consider the basic elements of this categorical geometric foundation in the study of perturbative renormalization theory.

5.1 Geometric nature of counterterms: Category of flat equi-singular connections

Reformulation of the Birkhoff decomposition in terms of classes of differential equations determines some geometric objects with respect to physical information of renormalizable theories. In fact, negative parts of decomposition of loops (or characters) can be applied to correct the behavior of solutions of these differential systems by flat connections (together with a special singularity, namely equi-singularity) near the singularities, without making more singularities elsewhere. The importance of this possibility can be investigated in making a geometrically encoding from counterterms (divergences) [14, 15, 19, 20, 21, 22]. In this part an overview from this story is done.

Definition 5.1.1. *For a given connected graded commutative Hopf algebra H of finite type with the related complex Lie group $G(\mathbb{C})$ and Lie algebra $\mathfrak{g}(\mathbb{C})$, let $\alpha : I = [a,b] \subset \mathbb{R} \longrightarrow \mathfrak{g}(\mathbb{C})$ be a smooth curve. Its associated time ordered exponential is defined by*

$$Te^{\int_a^b \alpha(t)dt} := 1 + \sum_{n \geq 1} \int_{a \leq s_1 \leq ... \leq s_n \leq b} \alpha(s_1)...\alpha(s_n)ds_1...ds_n$$

such that the product is taken in the graded dual space H^* and $1 \in H^*$ is the unit corresponding to the counit ϵ.

Remark 5.1.2. *(i) This integral only depends on 1-form $\alpha(t)dt$ and because of finite type property of Hopf algebra, for each element in H, it is finite with values in the affine group scheme.*
(ii) Time ordered exponential is the value $g(b)$ of the unique solution $g(t) \in G(\mathbb{C})$ of the differential equation

$$dg(t) = g(t)\alpha(t)dt$$

with the initial condition $g(a) = 1$.
(iii) It is multiplicative over the sum of paths.
(iv) Let ω be a flat $\mathfrak{g}(\mathbb{C})$-valued connection on $\Omega \subset \mathbb{R}^2$ and $\alpha : [0,1] \subset \mathbb{R} \longrightarrow \Omega$ be a curve. Then the time ordered exponential $Te^{\int_0^1 \alpha^*\omega}$ only depends on the homotopy class $[\alpha]$ of paths such that $\alpha(0) = a$, $\alpha(1) = b$.
(v) Let $\omega \in \mathfrak{g}(\mathbb{C}(\{z\}))$ has a trivial monodromy $M(\omega) = 1$. Then there exists a solution $g \in G(\mathbb{C}(\{z\}))$ for the equation $\mathbf{D}(g) = \omega$ such that the logarithmic derivative $\mathbf{D} : G(K) \longrightarrow \Omega^1(\mathfrak{g})$ is given by $\mathbf{D}(g) := g^{-1}dg$. [18, 19, 20, 21, 22]

Time ordered exponential necessarily and sufficiently provides an useful representation from loops on the set $Loop(G(\mathbb{C}), \mu)$ and their Birkhoff components such that it contains a reformulation of loops in terms of a class of differential systems associated to a family of equi-singular flat connections. As the consequence, this machinery delivers us a geometric meaningful from divergences.

Theorem 5.1.1. *Let $\gamma_\mu(z)$ be a loop in the class $Loop(G(\mathbb{C}), \mu)$. Then*
(i) There exists a unique $\beta \in \mathfrak{g}(\mathbb{C})$ and a loop $\gamma_{reg}(z)$ regular at $z = 0$ such that

$$\gamma_\mu(z) = Te^{-\frac{1}{z}\int_\infty^{-zlog\mu} \theta_{-t}(\beta)dt} \theta_{zlog\mu}(\gamma_{reg}(z)).$$

(ii) There is a representation from components of the Birkhoff factorization of $\gamma_\mu(z)$ based on time ordered exponential. We have

$$\gamma_{\mu+}(z) = Te^{-\frac{1}{z}\int_0^{-zlog\mu} \theta_{-t}(\beta)dt} \theta_{zlog\mu}(\gamma_{reg}(z)),$$

$$\gamma_-(z) = Te^{-\frac{1}{z}\int_0^\infty \theta_{-t}(\beta)dt}.$$

(iii) For each element $\beta \in \mathfrak{g}(\mathbb{C})$ and regular loop $\gamma_{reg}(z)$, one unique element of $Loop(G(\mathbb{C}), \mu)$ is identified. [18, 19, 20, 21, 22]

69

Definition 5.1.3. *Define an equivalence relation on connections (i.e. $\mathfrak{g}(\mathbb{C})$-valued one forms). ω_1, ω_2 are equivalent if there exists $h \in G(\mathbb{C}\{z\})$ such that*

$$\omega_2 = \mathbf{D}h + h^{-1}\omega_1 h.$$

Remark 5.1.4. *There is a correspondence between two equivalence $\mathfrak{g}(\mathbb{C})$-valued connections ω_1, ω_2 with trivial monodromies and negative parts of the Birkhoff decomposition of solutions of the equations*

$$\mathbf{D}f_1 = \omega_1, \quad \mathbf{D}f_2 = \omega_2$$

where $f_1, f_2 \in G(\mathbb{C}(\{z\}))$. It means that

$$\omega_1 \sim \omega_2 \iff f_1^- = f_2^-.$$

Dimensional regularization is an usual regularization technique in perturbative renormalization. It is based on an analytic continuation of Feynman integrals to the complex dimension $d \in \mathbb{C}$ in a neighborhood of the integral (critical) dimension D at which ultra-violet divergences occur. Under this regularization prescription, the procedure of renormalization with a special renormalization scheme namely, *minimal subtraction* (i.e. the subtraction of singular part of the Laurent series in $z = d - $ D at each order in the loop expansion) can be performed. In the BPHZ renormalization, all of the (sub-)divergences, in a recursive procedure, disappear such that it can be understood in terms of the extraction of finite values (i.e. Birkhoff decomposition of loops with values in the space of diffeographisms).

Theorem 5.1.2. *There is a principal bundle connected with a given theory under the Dim. Reg. + Min. Sub. scheme such that one special class of its related connections provides a geometric reinterpretation from physical information. This bundle is called renormalization bundle. [19, 20, 21, 22]*

Proof. A sketch of proof: Let Δ be an infinitesimal disk corresponds with the complexified dimension $D - z \in \Delta$ of dimensional regularization and $\mathbb{G}_m(\mathbb{C}) = \mathbb{C}^*$ be the possible choices for the normalization of an integral in dimension $D - z$. For the principal $\mathbb{C}^*$-bundle $\mathbb{G}_m \longrightarrow B \longrightarrow \Delta$, let $P = B \times G$ be a trivial principal G-bundle. Set

$$V := p^{-1}(\{0\}) \subset B, \quad B^0 := B - V, \quad P^0 = B^0 \times G.$$

This bundle, together with connections on it, can store some geometrical meanings related to physical information. □

Remark 5.1.5. *The choice of the unit of mass μ is the same as the choice of a section $\sigma : \Delta \longrightarrow B$ and indeed, the concept of equi-singularity turns to this fact.*

Definition 5.1.6. *Fix a base point $y_0 \in V$ and identify B with $\Delta \times \mathbb{G}_m(\mathbb{C})$. A flat connection ω on P^0 is called equi-singular, if*
- *It is $\mathbb{G}_m(\mathbb{C})$-invariant,*
- *For any solution f of the equation $\mathbf{D}f = \omega$, the restrictions of f to the sections $\sigma : \Delta \longrightarrow B$ with $\sigma(0) = y_0$ have the same singularity (namely, the pullbacks of a solution have the same negative parts of the Birkhoff decomposition, independent of the choice of the section and therefore the mass parameter).*

It is remarkable that one can expand equivalence relation given by definition 5.1.3 to this kind of connections. With working on classes of equi-singular flat connections on renormalization principal bundle, one can redisplay infinitesimal characters.

Theorem 5.1.3. *Equivalence classes of flat equi-singular $G-$connections on P^0 are represented by elements of $\mathfrak{g}(\mathbb{C})$ and also, each element of the Lie algebra of affine group scheme of the Hopf algebra associated to the renormalizable theory Φ identifies a specific class of equi-singular connections. The above process is independent of the choice of a local regular section $\sigma : \Delta \longrightarrow B$ with $\sigma(0) = y_0$. [19, 20, 21, 22]*

Theorems 5.1.1, 5.1.3 provide a bijective correspondence between negative parts of the Birkhoff decomposition of loops with values in $G(\mathbb{C})$ and classes of equi-singular flat connections on the renormalization bundle (which stores the regularization parameter). Since counterterms of the theory are identified with these negative parts (i.e. Connes-Kreimer formalization), therefore these classes of connections are presenting the counterterms.

In a categorical configuration, one can introduce a category such that equi-singular flat connections are its objects. The construction of this category and its properties are completely analyzed by Connes and Marcolli and just we review their main result.

Theorem 5.1.4. *For a given renormalizable QFT Φ underlying Dim. Reg. + Min. Sub. with the associated Lie group $G(\mathbb{C})$ and vector bundle $B \longrightarrow \Delta$, flat equi-singular connections on this bundle introduce an abelian tensor category with a specific fiber functor. Additionally, this category is a neutral Tannakian category and therefore it is equivalent to the category of finite dimensional representations of affine group scheme $G^* := G \rtimes \mathbb{G}_m$. [18, 19, 20, 21, 22]*

5.2 The construction of a universal Tannakian category

The Riemann-Hilbert correspondence conceptually consists of describing a certain category of equivalence classes of differential systems though a representation theoretic datum. For a given renormalizable QFT Φ with the related Hopf algebra H and the space of diffeographisms $G(\mathbb{C})$, it was shown that how one can identify a category of classes of flat equi-singular $G-$connections. Categorification of the Connes-Kreimer theory allows us to have a reasonable idea to formulate this story in a universal setting (underlying the Riemann-Hilbert problem) by constructing the universal category $\mathcal{E}$ of equivalence classes of all flat equi-singular vector bundles.

This category has the ability of covering the corresponding categories of all renormalizable theories and it means that when we are working on the theory Φ, it is possible to consider the subcategory $\mathcal{E}^\Phi$ of those flat equi-singular vector bundles which is equivalent to the finite dimensional linear representations of G^*. In this part we try to consider some general features of this very special category and its universality.

Start with a filtered vector bundle (E, W) over B with an increasing filtration

$$W^{-n-1}(E) \subset W^{-n}(E), \quad (W^{-n}(E) = \bigoplus_{m \geq n} E_m) \tag{5.2.1}$$

and set

$$Gr_n^W = \frac{W^{-n}(E)}{W^{-n-1}(E)}. \tag{5.2.2}$$

Definition 5.2.1. *A $W-$connection on E is a connection ∇ on the vector bundle $E^0 = E|_{B^0}$ such that*
- *∇ is compatible with the filtration,*
- *∇ is the trivial connection on Gr_n^W.*

Remark 5.2.2. *One can extend the mentioned equivalence relation in the definition 5.1.3 and the concept of equi-singularity to the level of $W-$connections. It means that*
(i) Two $W-$connections ∇_1, ∇_2 on E^0 are $W-$equivalent if there exists an automorphism T of E that preserves the filtration, where is identity on Gr_n^W and $T \circ \nabla_1 = \nabla_2 \circ T$.
(ii) A flat $W-$connection ∇ on E is equi-singular if it is $\mathbb{G}_m(\mathbb{C})-$invariant and the pullback along different sections σ of B (such that $\sigma(0) = y_0$) of a solution of the equation $\nabla \eta = 0$ have the same type of singularity.

Definition 5.2.3. *The couple (E, ∇) is called a flat equi-singular vector bundle.*

These pairs introduce a category $\mathcal{E}$ such that its objects $Obj(\mathcal{E})$ are data $\Theta = [V, \nabla]$ where V is a $\mathbb{Z}-$graded finite dimensional vector space and ∇ is an equi-singular $W-$connection on the filtered bundle $E^0 = B^0 \times V$ (with attention to the classes of connections).

Each morphism $T \in Hom(\Theta, \Theta')$ is a linear map $T : E \longrightarrow E'$ compatible with the grading such that it should have the following relation with the connections ∇, ∇'. Set

$$\nabla_1 := \begin{pmatrix} \nabla' & 0 \\ 0 & \nabla \end{pmatrix}, \quad \nabla_2 := \begin{pmatrix} \nabla' & T \circ \nabla - \nabla' \circ T \\ 0 & \nabla \end{pmatrix}. \tag{5.2.3}$$

For the defined connections in (5.2.3) on the vector bundle $(E' \bigoplus E)^*$, ∇_2 should be a conjugate of ∇_1 by the unipotent matrix $\begin{pmatrix} 1 & T \\ 0 & 1 \end{pmatrix}$.

Theorem 5.2.1. *With applying Riemann-Hilbert correspondence, one can formulate Connes-Marcolli category (consisting of divergences of a renormalizable theory) in a universal configuration by constructing the universal category $\mathcal{E}$ of equivalence classes of all flat equi-singular vector bundles with the the fiber functor given by $\varphi : \mathcal{E} \longrightarrow \mathcal{V}_{\mathbb{C}}, \Theta \longmapsto V$. [19, 20, 21, 22, 80]*

We know that there is a representation of a neutral Tannakian category by the category $\mathfrak{R}_{G^*}$ of finite dimensional representations of the affine group scheme of automorphisms of the fiber functor of the main category. This fact provides a new reformulation from this specific universal category.

Theorem 5.2.2. *$\mathcal{E}$ is a neutral Tannakian category. It is equivalent to the category $\mathfrak{R}_{\mathbb{U}^*}$ of finite linear representations of one special affine group scheme (related to the universal Hopf algebra of renormalization $H_{\mathbb{U}}$) namely, universal affine group scheme $\mathbb{U}^*$ such that $H_{\mathbb{U}}$ is a connected graded commutative non-cocommutative Hopf algebra of finite type. It is the graded dual of the universal enveloping algebra of the free graded Lie algebra $L_{\mathbb{U}} := \mathbf{F}(1, 2, ...)_{\bullet}$ generated by elements e_{-n} of degree $n > 0$ (i.e. one generator in each degree). [20, 22]*

Remark 5.2.4. *(i) Since the category of filtered vector spaces is not an abelian category, it is necessary to use the direct sum of bundles and the above condition on connections.*
(ii) The construction of this category over the field $\mathbb{Q}$ is also possible.

With attention to theorem 5.1.3 and also given correspondence in theorem 5.2.2, one can obtain a new prescription from equi-singular connections such that it will be applied to produce a new universal level for counterterms, renormalization groups and β−functions of theories.

Theorem 5.2.3. *Let H be a connected graded commutative Hopf algebra of finite type with the affine group scheme G and $\tilde{P}^0 := B^0 \times G^*$. Each equivalence class ω of flat equi-singular connections on $\tilde{P}^0$ identifies a graded representation $\rho_\omega : \mathbb{U}^* \longrightarrow G^*$ (which is identity on $\mathbb{G}_m$) and also, for each graded representation $\rho : \mathbb{U} \longrightarrow G$, there is one specific class of flat equi-singular connections. [19, 21]*

Based on the correspondence $\mathcal{R}_{\mathbb{U}^*} \simeq \mathcal{E}$, for each object $\Theta = [V, \bigtriangledown]$, there exists a unique representation ξ_Θ of $\mathbb{U}^*$ in V such that $\mathbf{D}\xi_\Theta(\gamma_\mathbb{U}) \simeq \bigtriangledown$. And also each arbitrary representation ξ of $\mathbb{U}^*$ in V determines a unique connection $\bigtriangledown$ (up to equivalence) such that $[V, \bigtriangledown]$ is an object in $\mathcal{E}$. It can be seen that one specific element will be determined from the process namely, the loop *universal singular frame* $\gamma_\mathbb{U}$ with values in $\mathbb{U}(\mathbb{C})$ where at this level one hopes to eliminate divergences for generating a finite theory.

Now because of the equivalence relation between loops (with values in the Lie group $G(\mathbb{C})$) and elements of the Lie algebra (corresponding to $G(\mathbb{C})$), with the help of a suitable element of the Lie algebra $L_\mathbb{U}$, one can characterize the universal singular frame.

Lemma 5.2.5. *The Lie algebra element corresponding to the universal singular frame is $e = \sum_{n\geq 1} e_{-n}$ (i.e. the sum of the generators of the Lie algebra).*

Since the universal Hopf algebra of renormalization is finite type, whenever we pair e with an element of the Hopf algebra, it will be only a finite sum. Hence e makes sense.

Theorem 5.2.4. *(i) e is an element in the completion of $L_\mathbb{U}$.*
(ii) $e : H_\mathbb{U} \longrightarrow \mathbb{K}[t]$ is a linear map. Its affine group scheme level namely, $\mathbf{rg} : \mathbb{G}_a(\mathbb{C}) \longrightarrow \mathbb{U}(\mathbb{C})$ is a morphism that plays an essential role to calculate the renormalization group.
(iii) The universal singular frame can be reformulated with $\gamma_\mathbb{U}(z, v) = Te^{-\frac{1}{z}\int_0^v u^Y(e)\frac{du}{u}}$.
(iv) For each loop $\gamma(z)$ in $Loop(G(\mathbb{C}), \mu)$, with help of the associated representation $\rho : \mathbb{U} \longrightarrow G$, the universal singular frame $\gamma_\mathbb{U}$ maps to the negative part $\gamma_-(z)$ of the Birkhoff decomposition of $\gamma(z)$ and also, the renormalization group $\{F_t\}_t$ in $G(\mathbb{C})$ can be recalculated by $\rho \circ \mathbf{rg}$. [19, 21]

When we apply the universal singular frame in the dimensional regularization, all divergences will be removed and it means that one can obtain a finite theory which depends only upon the choice of local trivialization of the $\mathbb{G}_m$−principal bundle B.

Positive and negative components of the Birkhoff factorization of the loop $\gamma_\mathbb{U}$ in the pro-unipotent affine group scheme $\mathbb{U}$ contain a universal meaning. For a given renormalizable theory Φ these components, via the identified representations in the theorem 5.2.3, map to renormalized values and counterterms, respectively. This fact provides a valuable concept namely, *universal counterterms*.

At last, it is favorable to emphasize the universality of the category $\mathcal{E}$ among other categories determined from physical theories. It means that this category gives us the power of analyzing flat equi-singular connections for each affine group scheme and this is the main reason of its universality.

Theorem 5.2.5. *Let H be a connected graded commutative Hopf algebra of finite type with the affine group scheme G. Let ω be a flat equi-singular connection on P^0 and $\psi : G \longrightarrow GL(V)$ a finite dimensional linear graded representation of G. We can correspond an element $\Theta \in Obj(\mathcal{E})$ to the data (ω, ψ). The equivalence class of ω identifies the same element Θ. [19, 20, 22]*

Remark 5.2.6. *For each flat equi-singular vector bundle $(E, \triangledown)$ (such that $E = B \times V$), the connection $\triangledown$ identifies a flat equi-singular G_V−valued connection ω. Indeed, $\triangledown$ is given by adding a Lie G_V−valued one form to the trivial connection. This correspondence preserves the mentioned equivalence relation in the definition 5.1.3.*

In summary, it is observed that lifting the Connes-Kreiemr perturbative renormalization to the universal configuration can yield the particular Hopf algebra H_U. Because of the combinatorial nature of the universal Connes-Kreimer Hopf algebra, it should be reasonable to search a hidden combinatorial construction in the backbone of this Hopf algebra. This problematic notion is the main topic of the next chapter and its importance will be investigated when we want to generalize the Connes-Marcolli treatment in the study of Dyson-Schwinger equations.

Chapter 6

Universal Hopf Algebra of Renormalization

We discussed that how one can systematically interpret the hidden combinatorics of perturbative renormalization based on the Hopf algebra structure on Feynman diagrams of a pQFT. Furthermore, it was exhibited that the Connes-Kreimer Hopf algebra of rooted trees (equipped with some decorations which represent primitive 1PI graphs) plays the role of an available practical model such that with changing labels, it will upgrade for each arbitrary theory. On the other hand, we saw that the universal affine group scheme $\mathbb{U}$ governs the structure of divergences of all renormalizable theories and the universality of $H_\mathbb{U}$ turns to its independency from all theories.

In this chapter, we are going to discover the natural combinatorial backbone of $H_\mathbb{U}$ by giving a new explicit rooted tree type reformulation from this particular Hopf algebra and its related Lie group. Then with using this new interpretation, first we obtain rigorous relations between universal Hopf algebra of renormalization and some well-known combinatorial Hopf algebras. Secondly, we expand this new Hall tree formalism to the level of Lie groups and finally, we will describe the universal singular frame based on Hall polynomials such that as the result, new Hall tree scattering formulaes for physical information will be determined.

6.1 Shuffle type representation

The Hopf algebra $H_\mathbb{U}$ is introduced in the theorem 5.2.2 and in particular, one important note is that as an algebra it is isomorphic to the linear space of noncommutative polynomials in variables f_n, $n \in \mathbb{N}_{>0}$ with the shuffle product. It is a skeleton key for us to find a relation between this Hopf algebra and rooted trees and because of that at first we need some more information about shuffle structures.

Definition 6.1.1. *Let V be a vector space over the field $\mathbb{K}$ of characteristic zero and $T(V) = \bigoplus_{n \geq 0} V^{\otimes n}$ be its related tensor algebra. Set*

$$S(m,n) = \{\sigma \in S_{m+n} : \sigma^{-1}(1) < ... < \sigma^{-1}(m), \quad \sigma^{-1}(m+1) < ... < \sigma^{-1}(m+n)\}.$$

It is called the set of (m,n)–shuffles. For each $x = x_1 \otimes ... \otimes x_m \in V^{\otimes m}$, $y = y_1 \otimes ... \otimes y_n \in V^{\otimes n}$

and $\sigma \in S(m,n)$, define

$$\sigma(x \otimes y) = u_{\sigma(1)} \otimes u_{\sigma(2)} \otimes ... \otimes u_{\sigma(m+n)} \in V^{\otimes(m+n)}$$

such that $u_k = x_k$ for $1 \leq k \leq m$ and $u_k = y_{k-m}$ for $m+1 \leq k \leq m+n$. The shuffle product of x, y is given by

$$x \star y := \sum_{\sigma \in S(m,n)} \sigma(x \otimes y).$$

Lemma 6.1.2. $(T(V), \star)$ *is a unital commutative associative algebra. [26, 42, 88]*

There are some extensions of this product such as quasi-shuffles, mixable shuffles. Let A be a *locally finite set* (i.e. a disjoint union of finite sets $A_n, n \geq 1$). The elements of A are *letters* and monomials are called *words* such that the empty word is denoted by 1. Set $A^- := A \cup \{0\}$.

Definition 6.1.3. *Function* $<.,.>: A^- \times A^- \longrightarrow A^-$ *is called a Hoffman pairing, if it satisfies following conditions:*
- *For all* $a \in A^-$, $<a, 0>= 0$,
- *For all* $a, b \in A^-$, $<a, b>=<b, a>$,
- *For all* $a, b, c \in A^-$, $<<a, b>, c>=<a, <b, c>>$,
- *For all* $a, b \in A^-$, $<a, b>= 0$ *or* $|<a, b>| = |a| + |b|$.

A locally finite set A together with a Hoffman pairing $<.,.>$ on A^- is called a Hoffman set.

Definition 6.1.4. *Let* $\mathbb{K} < A >$ *be the graded noncommutative polynomial algebra over* $\mathbb{K}$. *The quasi-shuffle product* $\star^-$ *on* $\mathbb{K} < A >$ *is defined recursively such that for any word* w, $1 \star^- w = w \star^- 1 = w$ *and also for words* w_1, w_2 *and letters* a, b,

$$(aw_1) \star^- (bw_2) = a(w_1 \star^- bw_2) + b(aw_1 \star^- w_2) + <a, b> (w_1 \star^- w_2).$$

Lemma 6.1.5. $\mathbb{K} < A >$ *together with the product* $\star^-$ *is a graded commutative algebra such that when* $<.,.>= 0$, *it will be the shuffle algebra* $(T(V), \star)$ *where V is a vector space generated by the set A. [26, 42, 88]*

Shuffle type products can determine interesting family of Hopf algebras such that it can be possible to reformulate H_U on the basis of this class. The next theorem gives a complete characterization from this class of Hopf algebras.

Theorem 6.1.1. *(i) The (quasi-) shuffle product introduces a graded connected commutative noncocommutative Hopf algebra structure (of finite type) on $(\mathbb{K} < A >, \star^-)$ and $(T(V), \star)$.*
(ii) There is an isomorphism (as a graded Hopf algebras) between $(T(V), \star)$ and $(\mathbb{K} < A >, \star^-)$.
(iii) There is a graded connected Hopf algebra structure (of finite type) (comes from (quasi-)shuffle product) on the graded dual of $\mathbb{K} < A >$.
(iv) We can extend the isomorphism in the second part to the graded dual level. [26, 42, 88]

Proof. The compatible Hopf algebra structure on the shuffle algebra of noncommutative polynomials is given by the coproduct

$$\Delta(w) = \sum_{uv=w} u \otimes v$$

and the counit

$$\epsilon(1) = 1, \quad \epsilon(w) = 0, \ w \neq 1.$$

For a given Hoffman pairing $< .,. >$ and any finite sequence S of elements of the set A, with induction define $< S > \in A^-$ such that for any $a \in A$,

$$< a > = a, \quad < a, S > = < a, < S >> .$$

Let $C(n)$ be the set of compositions of n and $C(n,k)$ be the set of compositions of n with length k. For each word $w = a_1...a_n$ and composition $I = (i_1, ..., i_l)$, set

$$I < w > := < a_1, ..., a_{i_1} > < a_{i_1+1}, ..., a_{i_1+i_2} > ... < a_{i_1+...+i_{l-1}+1}, ..., a_n > .$$

It means that compositions act on words. Now for any word $w = a_1...a_n$, its antipode is given by

$$S(1) = 1,$$

$$S(w) = -\sum_{k=0}^{n-1} S(a_1...a_k) \star^- a_{k+1}...a_n = (-1)^n \sum_{I \in C(n)} I < a_n...a_1 > .$$

The isomorphism between these Hopf algebra structures (compatible with the shuffle products) is given by morphisms

$$\tau(w) = \sum_{(i_1,...,i_l) \in C(|w|)} \frac{1}{i_1!...i_l!} (i_1, ..., i_l) < w >,$$

$$\psi(w) = \sum_{(i_1,...,i_l) \in C(|w|)} \frac{(-1)^{|w|-l}}{i_1...i_l} (i_1, ..., i_l) < w > .$$

The graded dual $\mathbb{K} < A >^*$ has a basis consisting of elements v^* (where v is a word on A) with the following pairing such that if $u = v$, then $(u, v^*) = 1$ and if $u \neq v$, then $(u, v^*) = 0$. Its Hopf algebra structure is given by the concatenation product

$$conc(u^* \otimes v^*) = (uv)^*$$

and the coproduct

$$\delta(w^*) = \sum_{u,v} (u \star^- v, w^*) u^* \otimes v^*.$$

The map

$$\tau^*(u^*) = \sum_{n \geq 1} \sum_{<a_1,...,a_n> = u} \frac{1}{n!} (a_1...a_n)^*,$$

is an isomorphism in the dual level and its inverse is given by

$$\psi^*(u^*) = \sum_{n \geq 1} \frac{(-1)^{n-1}}{n} \sum_{<a_1,...,a_n> = u} (a_1...a_n)^*.$$

□

Because of the relation between Hopf algebras and Lie theory, one can consider shuffle Hopf algebras at the Lie algebra version.

Definition 6.1.6. *Let $\mathcal{L}$ be a Lie algebra over $\mathbb{K}$. There exists an associative algebra $\mathcal{L}_0$ over $\mathbb{K}$ together with a Lie algebra homomorphism $\phi_0 : \mathcal{L} \longrightarrow \mathcal{L}_0$ such that for each couple $(\mathcal{A}, \phi : \mathcal{L} \longrightarrow \mathcal{A})$ of an algebra and a Lie algebra homomorphism, there is a unique algebra homomorphism $\phi_{\mathcal{A}} : \mathcal{L}_0 \longrightarrow \mathcal{A}$ such that $\phi_{\mathcal{A}} \circ \phi_0 = \phi$. $\mathcal{L}_0$ is called universal enveloping algebra of $\mathcal{L}$ and it is unique up to isomorphism.*

Lemma 6.1.7. *(i) Universal enveloping algebra $\mathcal{L}_0$ of the free Lie algebra $\mathcal{L}(A)$ is a free associative algebra on A.*
(ii) ϕ_0 is an injective morphism such that $\phi_0(\mathcal{L}(A))$ is the Lie subalgebra of $\mathcal{L}_0$ generated by $\phi_0 \circ i(A)$. [88]

Definition 6.1.8. *The set of Lie polynomials in $\mathbb{K} < A >^*$ is the smallest sub-vector space of $\mathbb{K} < A >^*$ containing the set of generators $A^* := \{a^* : a \in A\}$ and closed under the Lie bracket.*

Corollary 6.1.9. *(i) The set of Lie polynomials in $\mathbb{K} < A >^*$ forms a Lie algebra. It is the free Lie algebra on A^* such that $\mathbb{K} < A >^*$ is its universal enveloping algebra.*
(ii) In the shuffle product case, the Lie polynomials are exactly the primitives for δ and therefore at the level of (quasi-)shuffle product, the primitives are elements of the form $\psi^* p$ such that p is a Lie polynomial [42].

Now it is the place to come back to the definition of the universal Hopf algebra of renormalization. It can be seen that the set $A = \{f_n : n \in \mathbb{N}_{>0}\}$ is a locally finite set and as an algebra, H_U is isomorphic to $(T(V), \star)$ such that V is a vector space over $\mathbb{C}$ spanned by the set A. Therefore its Hopf algebra structure is determined with the theorem 6.1.1. At the Lie algebra level, we have to go to the dual structure. Corollary 6.1.9 shows us that the set of all Lie polynomials in H_U^* is the free Lie algebra generated by $\{f_n^*\}_{n \in \mathbb{N}_{>0}}$ such that H_U^* is its universal enveloping algebra and on the other hand, we know that H_U^* is identified by the universal enveloping of the free graded Lie algebra L_U generated by $\{e_{-n}\}_{n \in \mathbb{N}_{>0}}$.

Remark 6.1.10. *This procedure implies to have a reasonable correspondence between generators of the Lie algebra L_U and elements of the set A^*.*

6.2 Rooted tree type representation

In this part, with attention to the shuffle nature of H_U, we introduce a combinatorial version from this specific Hopf algebra in the Connes-Marcolli universal renormalization theory. We consider completely a new Hall rooted tree reformulation from H_U to obtain interesting relations between this particular Hopf algebra and some well-known combinatorial Hopf algebras introduced in [7, 17, 42, 43, 44, 45, 46, 47, 74, 75, 87, 92, 100]. Moreover, we extend this rooted tree representation of H_U to the level of its associated Lie group to display its elements based on formal series of Hall trees. So it provides a new Hall tree type representation from universal singular frame γ_U where we will consider the applications of Hall basis and PBW basis to reformulate combinatorially physical information.

Definition 6.2.1. *Defining a partial order $\preceq$ on the set of all rooted trees $\mathbf{T}$. We say $t \preceq s$, if t can be obtained from s by removing some non-root vertices and edges and it implies that $|t| \leq |s|$.*

Definition 6.2.2. Let $\mathbf{T(A)}$ ($\mathbf{F(A)}$) be the set of all rooted trees (forests) labeled by the set A.
(i) For $a \in A$, $t_1, ..., t_m \in \mathbf{T(A)}$ such that $u = t_1...t_m \in \mathbf{F(A)}$, $B_a^+(u)$ is a labeled rooted tree of degree $|t_1| + ... + |t_m| + 1$ obtained by grafting the roots of $t_1, ..., t_m$ to a new root labeled by a. In addition, $B_a^+(\mathbb{I})$ is a rooted tree with just one labeled vertex.
(ii) For $t \in \mathbf{T(A)}$ and $u \in \mathbf{F(A)}$, define a new element $t \circ u$ such that it is a labeled rooted tree of degree $|t| + |u|$ given by grafting the roots of labeled rooted trees in u to the root of t.

Lemma 6.2.3. (i) The operation $\circ$ is not associative.
(ii) $\forall t \in \mathbf{T(A)}$, $\forall u, v \in \mathbf{F(A)} : (t \circ u) \circ v = t \circ (uv) = (t \circ v) \circ u$.
(iii) $t_1 \circ ... \circ t_m \circ u = t_1 \circ (t_2 \circ ... \circ (t_m \circ u))$, $t^{\circ k} = t \circ ... \circ t$, k times.
(iv) For each $u \in \mathbf{F(A)}$, let $per(u)$ be the number of different permutations of the vertices of a labeled partially ordered set that representing u. Then

$$per(\mathbb{I}) = 1, \quad per(B_a^+(u)) = per(u).$$

And if $u = \prod_{j=1}^{m}(t_j)^{i_j}$, then

$$per(u) = \prod_{j=1}^{m} i_j! per(t_j)^{i_j}.$$

(v) The bilinear extension of $\circ$ to the linear combinations of labeled rooted trees (or linear combinations of labeled forests) is also possible.

Definition 6.2.4. A set $\mathbf{H(T(A))}$ of labeled rooted trees is called Hall set, if it has following conditions:
- There is a total order relation $>$ on $\mathbf{H(T(A))}$.
- If $a \in A$, then $B_a^+(\mathbb{I}) \in \mathbf{H(T(A))}$.
- For $a \in A$, $u \in \mathbf{F(A)} - \{\mathbb{I}\}$ such that $u = t_1^{\circ r_1}...t_m^{\circ r_m}$, $t_1, ..., t_m \in \mathbf{H(T(A))}$, $r_1, ..., r_m \geq 1$, $t_1 > ... > t_m$,

$$B_a^+(u) \in \mathbf{H(T(A))} \iff t_m > B_a^+(t_1^{\circ r_1}...t_{m-1}^{\circ r_{m-1}}) \in \mathbf{H(T(A))}.$$

- If $t = B_a^+(t_1^{\circ r_1}...t_m^{\circ r_m}) \in \mathbf{H(T(A))}$ such that $t_1, ..., t_m \in \mathbf{H(T(A))}$, $r_1, ..., r_m \geq 1$, $a \in A$, then for each $j = 1, ..., m$, $t_j > t$.

Lemma 6.2.5. For each $t \in \mathbf{H(T(A))}$, $r \geq 1$ and $a \in A$, it is easy to see that

$$B_a^+(t^{\circ r}) \in \mathbf{H(T(A))} \iff t > B_a^+(\mathbb{I}).$$

Definition 6.2.6. For a Hall set $\mathbf{H(T(A))}$, the set of its forests is given by

$$\mathbf{H(F(A))} := \{\mathbb{I}\} \cup \{t_1^{r_1}...t_m^{r_m} : r_1, ..., r_m \geq 1, t_1, ..., t_m \in \mathbf{H(T(A))}, t_i \neq t_j (i \neq j)\}.$$

Lemma 6.2.7. Elements of $\mathbf{H(F(A))}$ and rooted trees are in the relation with the map

$$\xi : \mathbf{H(F(A))} - \{\mathbb{I}\} \longrightarrow \mathbf{T(A)}$$

$$t_1^{\circ r_1}...t_m^{\circ r_m} \longrightarrow t_1^{\circ r_1} \circ (t_2^{\circ r_2}...t_m^{\circ r_m}).$$

ξ is injective and its image is the set $\{B_a^+(u) \in \mathbf{T(A)} : u \in \mathbf{H(F(A))}, a \in A\}$. [77]

Remark 6.2.8. *(i) Hall trees and Hall forests have no symmetry.*
(ii) There is a one to one correspondence between a Hall set of A−labeled rooted trees and a Hall set of words on A. [88]

Definition 6.2.9. *For $t \in \mathbf{H}(\mathbf{T}(\mathbf{A}))$, there is a standard decomposition $(t^1, t^2) \in \mathbf{H}(\mathbf{T}(\mathbf{A})) \times \mathbf{H}(\mathbf{T}(\mathbf{A}))$ such that*
- *If $|t| = 1$, then the decomposition is $t^1 = t$, $t^2 = \mathbb{I}$,*
- *And if $t = B_a^+(t_1^{or_1}...t_m^{or_m})$ such that $r_1, ..., r_m \geq 1$, $t_1, ..., t_m \in \mathbf{H}(\mathbf{T}(\mathbf{A})) : t_1 > ... > t_m$, $a \in A$, then the decomposition is given by*

$$t^1 = B_a^+(t_1^{or_1}...t_{m-1}^{or_{m-1}} t_m^{or_m-1}), \quad t^2 = t_m.$$

- *For a Hall forest $u \in \mathbf{H}(\mathbf{F}(\mathbf{A})) - \mathbf{H}(\mathbf{T}(\mathbf{A}))$ such that $u = t_1^{or_1}...t_m^{or_m}$, $t_1 > ... > t_m$, the decomposition is given by $(u^1, u^2) \in \mathbf{H}(\mathbf{F}(\mathbf{A})) \times \mathbf{H}(\mathbf{T}(\mathbf{A}))$ where*

$$u^1 = t_1^{or_1}...t_{m-1}^{or_{m-1}} t_m^{or_m-1}, \quad u^2 = t_m.$$

Definition 6.2.10. *For a given map that associates to each word w on A a scalar $\alpha_w \in \mathbb{K}$, define a map $\alpha : \mathbf{F}(\mathbf{A}) \longrightarrow \mathbb{K}$ such that*
- $\mathbb{I} \longmapsto \alpha_1$.
- *For each $u \in \mathbf{F}(\mathbf{A}) - \{\mathbb{I}\}$, there is a labeled partially ordered set $(\mathbf{u}(A), \geq)$ that represents the forest u such that vertices $x_1, ..., x_n, ...$ of this poset are labeled by $l(x_i) = a_i \in A$ $(1 \leq i)$. Let $>_{\mathbf{u}(A)}$ be a total order relation on the set of vertices $\mathbf{u}(A)$ such that it is an extension of the partial order relation $\geq$ on $\mathbf{u}(A)$. For each ordered sequence $x_{i_1} >_{\mathbf{u}(A)} ... >_{\mathbf{u}(A)} x_{i_n}$ in $\mathbf{u}(A)$, its corresponding word $a_{i_1}...a_{i_n}$ is denoted by $w(>_{\mathbf{u}(A)})$. Set*

$$\alpha(u) := \sum_{>_{\mathbf{u}(A)}} \alpha_{w(>_{\mathbf{u}(A)})},$$

where the sum is over all total order relations $>_{\mathbf{u}(A)}$ (i.e. extensions of the main partial order relation $\geq$) on the set of vertices of $\mathbf{u}(A)$.

Lemma 6.2.11. *Define a map π given by*

$$\pi : \mathbb{K}[\mathbf{T(A)}] \longrightarrow (\mathbb{K}<A>, \star^-), \quad \pi(u) := \sum_{>_{\mathbf{u}(A)}} w(>_{\mathbf{u}(A)}).$$

One can show that for each $u, v \in \mathbf{F(A)}$ and $a \in A$,
(i) $\pi(B_a^+(u)) = \pi(u)a$,
(ii) $\pi(uv) = \pi(u) \star^- \pi(v)$,
(iii) $\alpha(u) = \widehat{\alpha}(\pi(u))$, where $\widehat{\alpha} : (\mathbb{K}<A>, \star^-) \longrightarrow \mathbb{K}$, $\widehat{\alpha}(w) = \alpha_w$ is a $\mathbb{K}$−linear map.

Definition 6.2.12. *For given maps $\alpha, \beta : \mathbf{F(A)} \longrightarrow \mathbb{K}$, one can define a new map $\alpha\beta : \mathbf{F(A)} \longrightarrow \mathbb{K}$ given by*

$$u \longmapsto \sum_{(\mathfrak{v}(A), \mathfrak{w}(A)) \in R(\mathbf{u}(A))} \alpha(v)\beta(w)$$

such that $\mathbf{u}(A)$ is a labeled poset representing u and $\mathfrak{v}(A), \mathfrak{w}(A)$ are labeled partially ordered subsets of $\mathbf{u}(A)$ such that the set of all pairs $(\mathfrak{v}(A), \mathfrak{w}(A))$ with the following conditions is denoted by

80

$R(\mathfrak{u}(A))$.
- The set of vertices in $\mathfrak{u}(A)$ is the disjoint union of the set of vertices $\mathfrak{v}(A)$ and $\mathfrak{w}(A)$,
- For each $x, y \in \mathfrak{u}(A)$ such that $x \geq y$, if $x \in \mathfrak{w}(A)$ then $y \in \mathfrak{w}(A)$.

Remark 6.2.13. It can be seen that for each word $w = a_1...a_m$ on A,

$$(\alpha\beta)_w = \alpha_w\beta_1 + \alpha_1\beta_w + \sum_{j=1}^{m-1} \alpha_{a_1...a_j}\beta_{a_{j+1}...a_m}.$$

Definition 6.2.14. For the map α given by the definition 6.2.10, one can introduce an equivalence relation on $\mathbb{K}[T(A)]$ such that for each $u, v \in \mathbb{K}[T(A)]$, they are congruent $(u \equiv v)$, if for every map $\widehat{\alpha} : A^* \longrightarrow \mathbb{K}$, $w \mapsto \alpha_w$, (such that A^* is the set of all words on A), then we have $\alpha(u) = \alpha(v)$.

Remark 6.2.15. (i) $u \equiv v \iff u - v \in \ker \pi$.
(ii) $u, v \in \mathbb{K}[T(A)]$, $a \in A$, $u \equiv v \Longrightarrow B_a^+(u) \equiv B_a^+(v)$.
(iii) $\overline{u}, \overline{v} \in F(A) : \overline{u} \equiv \overline{v} \Longrightarrow u\overline{u} \equiv v\overline{v}$.
(iv) $t \in T(A), n \geq 1 : t^n \equiv n! t^{\circ n}$.
(v) $t \in T(A), i, j \geq 1 : t^{\circ i} t^{\circ j} \equiv \frac{(i+j)!}{i!j!} t^{\circ(i+j)}$.

Lemma 6.2.16. For $m \geq 2$ and $t_1, ..., t_m \in T(A)$, with induction one can show that

$$t_1...t_m \equiv \sum_{i=1}^{m} t_i \circ \prod_{j \neq i} t_j.$$

There is an algorithm (in finite number of recursion steps) for rewriting each $u \in F(A)$ as $u \equiv v$ such that $v \in \mathbb{K}\mathbf{H}(\mathbf{F}(A))$ (i.e. v is a $\mathbb{K}$−linear combination of Hall forests). [77]

Definition 6.2.17. There is a canonical map f on Hall rooted trees defined by
- $f(a) = a$, if $a \in A$,
- $f(t) = f(t^1)f(t^2)$, if t be of degree ≥ 2 with the standard decomposition $t = (t^1, t^2)$.
The function f is called foliage and for each Hall tree t, its degree $|f(t)|$ is the number of leaves of t. The foliage of a Hall tree is called Hall word.

Theorem 6.2.1. For each word w on A, there is a unique factorization $w = f(t_1)...f(t_n)$ such that $t_i \in \mathbf{H}(\mathbf{T}(\mathbf{A}))$ and $t_1 > ... > t_n$. [88]

One can show that Hall sets of A-labeled rooted trees can be reconstructed recursively from an arbitrary Hall set of words on A. It means that

Corollary 6.2.18. A Hall set of words on A is the image under the foliage of a Hall set $\mathbf{H}(\mathbf{T}(\mathbf{A}))$ of labeled rooted trees.

There is an important class of words (i.e. Lyndon words) such that one can deduce Hall sets from them and moreover, this kind of words can store interesting information about shuffle structures. Let us start with a well-known order.

Definition 6.2.19. Let A be a totally ordered set. The alphabetical ordering determines a total order on the set of words on A such that for any nonempty word v, put $u < uv$ and also for letters $a < b$ and words w_1, w_2, w_3, put $w_1 a w_2 < w_1 b w_3$.

Definition 6.2.20. *For a given total order set A, a non-trivial word w is called Lyndon, if for any non-trivial factorization $w = uv$, we have $w < v$.*

The first advantage of these words can be seen in their influencing role in making a Hall set.

Theorem 6.2.2. *The set of Lyndon words, ordered alphabetically, is a Hall set. [42, 88]*

Theorem 6.2.3. *Let A be a locally finite set equipped with a total order relation. The (quasi-)shuffle algebra $(\mathbb{K} < A >, \star^{-})$ is the free polynomial algebra on the Lyndon words. [42]*

It is shown that the universal Hopf algebra of renormalization as an algebra is defined by the shuffle product on the linear space of noncommutative polynomials with variables f_n ($n \in \mathbb{N}$). This determines an important order.

Definition 6.2.21. *With referring to given correspondence in the result 6.1.10, one can define a natural total order relation (depending on the degrees of the generators e_{-n}, ($n \in \mathbb{N}$) of the free Lie algebra L_U) on the set $A = \{f_n : n \in \mathbb{N}_{>0}\}$. It is given by*

$$f_m > f_n \iff n > m.$$

Theorem 6.2.3 shows that H_U (as an algebra) is the free polynomial algebra of Lyndon words on the set A and therefore one can consider Hall set of these Lyndon words (ordered alphabetically) such that its corresponding Hall set of labeled rooted trees is denoted by $\mathbf{H}(\mathbf{T}(\mathbf{A}))_U$. It can be seen obviously that the set of Lyndon words is an influencing factor in determining this bridge between rooted trees and H_U.

It is near to have our interesting rooted tree reformulation. Let us consider free commutative algebra $\mathbb{K}[\mathbf{T}(\mathbf{A})]$ such that the set $\{t_1^{r_1}...t_m^{r_m} : t_1,...,t_m \in \mathbf{T}(\mathbf{A})\}$ is a $\mathbb{K}$−basis (as a graded vector space) where each expression $t_1^{r_1}...t_m^{r_m}$ is a forest.

Definition 6.2.22. *For the forest u with the associated partial order set $(\mathfrak{u}(A), \geq)$, define a coproduct given by*

$$\Delta(u) = \sum_{(\mathfrak{v}(A),\mathfrak{w}(A)) \in R(\mathfrak{u}(A))} v \otimes w$$

such that labeled forests v, w are represented by labeled partially ordered subsets $\mathfrak{v}(A), \mathfrak{w}(A)$ of $\mathfrak{u}(A)$.

Theorem 6.2.4. *Coproduct 6.2.22 determines a connected graded commutative Hopf algebra structure on $\mathbb{K}[\mathbf{T(A)}]$ such that the product in the dual space $\mathbb{K}[\mathbf{T(A)}]^\star = \{\alpha : \mathbf{T(A)} \longrightarrow^{linear} \mathbb{K}\}$ corresponds to the fixed coproduct namely, dual of the coalgebra structure and it means that for each $\alpha, \beta \in \mathbb{K}[\mathbf{T(A)}]^\star$ and each forest u,*

$$\alpha\beta(u) = (\alpha \otimes \beta)\Delta(u).$$

[77]

Remark 6.2.23. *One can show that H_{GL} (labeled by the set A) and $\mathbb{K}[\mathbf{T(A)}]$ are graded dual to each other.*

By using theorem 6.2.4 and operation B_a^+, one can show that this Hopf algebra has a universal property.

Theorem 6.2.5. *Let H be a commutative Hopf algebra over $\mathbb{K}$ and $\{L_a : H \longrightarrow H\}_{a \in A}$ be a family of $\mathbb{K}$-linear maps such that*

$$\cup_{a \in A} Im L_a \subset ker \epsilon_H$$

and

$$\Delta_H L_a(c) = L_a(c) \otimes \mathbb{I}_H + (id_H \otimes L_a)\Delta_H(c).$$

Then there exists a unique Hopf algebra homomorphism $\psi_H : \mathbb{K}[T(A)] \longrightarrow H$ such that for each $u \in \mathbb{K}[T(A)]$ and $a \in A$, we have

$$\psi_H(B_a^+(u)) = L_a(\psi_H(u)).$$

[77]

Corollary 6.2.24. *Theorem 6.2.5 is a poset version of the combinatorial Connes-Kreimer Hopf algebra and it means that H_{CK} (labeled by the set A) is isomorphic to $\mathbb{K}[T(A)]$.*

Proof. It is clearly proved based on the theorem 2.2.2 (universal property of H_{CK}). □

Lemma 6.2.25. *There is a bijection between the set of non-empty words and the set of labeled rooted trees without side-branchings.*

Proof. One can show that the $\mathbb{K}$-linear map π is a Hopf algebra homomorphism and for each $a_1, ..., a_m \in A$, we have

$$\pi(B_{a_m}^+...B_{a_2}^+(a_1)) = a_1...a_m.$$

□

Corollary 6.2.26. *The map π is an epimorphism and for each $\widehat{\alpha}, \widehat{\beta} \in \mathbb{K} < A >^*$, $u \in \mathbb{K}[T(A)]$, we have*

$$< \widehat{\alpha}\widehat{\beta}, \pi(u) > = < \alpha\beta, u >$$

such that $\alpha, \beta \in \mathbb{K}[T(A)]^*$.

Everything is ready to introduce a rooted tree version of shuffle type Hopf algebras.

Theorem 6.2.6. *(i) The (quasi-)shuffle Hopf algebra $(\mathbb{K} < A >, \star^-)$ is isomorphic to the quotient Hopf algebra $\frac{\mathbb{K}[T(A)]}{I_\pi}$ such that $I_\pi := Ker\pi$ is a Hopf ideal in $\mathbb{K}[T(A)]$ with the generators $< \{\prod_{i=1}^m t_i - \sum_{i=1}^m t_i \circ \prod_{j \neq i} t_j : m > 1, t_1, ..., t_m \in T(A)\} >=< \{t \circ z + z \circ t - tz : t, z \in T(A)\} \cup \{s \circ t \circ z + s \circ z \circ t - s \circ (tz) : t, z, s \in T(A)\} >=< \{t \circ z + z \circ t - tz : t, z \in T(A)\} \cup \{s \circ (tz) + z \circ (ts) + t \circ (sz) - tzs : t, z, s \in T(A)\} >$.*
(ii) As an $\mathbb{K}$-algebra, $\frac{\mathbb{K}[T(A)]}{I_\pi}$ is freely generated by the set

$$\{t + I_\pi : t \in \mathbf{H}(\mathbf{T}(\mathbf{A}))\}.$$

[77]

With attention to the given operadic picture from Connes-Kreimer Hopf algebra in the past chapters, next result can be indicated immediately.

83

Lemma 6.2.27. *For each locally finite set A together with a total order relation, there exist Hopf ideals J_1, J_2 such that*

$$(\mathbb{K} < A >, \star^-) \cong \frac{\mathbb{K}[T(A)]}{I_\pi} \cong \frac{H_{CK}(A)}{J_1} \cong \frac{H_{NAP}(A)}{J_2}.$$

Proof. From theorems 2.2.9, 6.2.5, it is enough to set $J_1 := I_\pi$ and $J_2 := \rho^{-1} J_1$. □

Proposition 6.2.28. *Consider the incidence Hopf algebra with respect to the basic set operad NAP which is labeled by the set $A = \{f_n\}_n$. A quotient of this Hopf algebra determines isomorphically the universal Hopf algebra of renormalization.*

Proof. Theorem 6.2.6 provides a rooted tree representation of H_U. Now apply the lemma 6.2.27. Just it is enough to replace the set A with the variables f_n such that the identified Lyndon words (with the shuffle structure of H_U) gives us the Hall set $\mathbf{H}(\mathbf{T}(\mathbf{A}))_U$. □

In continue of this part, we use this new redefinition from universal Hopf algebra of renormalization to provide some interesting relations between this Hopf algebra and some well-known Hopf algebra structures.

Corollary 6.2.29. *Applying the obtained Hall rooted tree type representation of H_U determines the following commutative diagrams of Hopf algebra homomorphisms.*

$$\begin{array}{ccccccc}
NSYM & \xrightarrow{\beta_1} & H_F & & SYM & \xleftarrow{\beta_4^*} & U(L_U) \\
\beta_3 \downarrow & & \beta_2 \downarrow & & \beta_3^* \downarrow & & \beta_2^* \downarrow \\
SYM & \xrightarrow{\beta_4} & H_U & & QSYM & \xleftarrow{\beta_1^*} & H_P
\end{array} \quad (6.2.1)$$

Proof. We know that $SYM \subset QSYM$. As a vector space, $QSYM$ is generated by the monomial quasi-symmetric functions M_I such that $I = (i_1, ..., i_k)$ and

$$M_I := \sum_{n_1 < n_2 < ... < n_k} x_{n_1}^{i_1} ... x_{n_k}^{i_k}. \quad (6.2.2)$$

If we remove order in a composition, then the generators $m_\lambda := \sum_{\mathbf{c}(I) = \lambda} M_I$ of SYM (viewed as a vector space) can be obtained where the map $\mathbf{c}$ maps compositions to partitions that forgets order. The Hopf algebra structure on $QSYM$ is determined by the coproduct

$$\Delta(M_{(i_1,...,i_k)}) = \sum_{j=0}^{k} M_{(i_1,...,i_j)} \otimes M_{(i_{j+1},...,i_k)}. \quad (6.2.3)$$

It is easy to see that the power-sum symmetric functions M_i are primitives and elementary symmetric functions $M_{(1,...,1)}$ are divided powers. As an algebra, $NSYM$ is the noncommutative polynomials on the variables z_n of degree n such that by applying the generators z_n as divided powers, its related Hopf algebra structure can be determined. [41, 44, 45, 46]

For each (planar) rooted tree t, one can put different decorations (with elements of the locally finite set $A = \{f_n\}_{n \geq 1}$) on its vertices. Let $[t]$ be the class of all different possible Hall (planar) rooted trees with respect to t in $\mathbf{H}(\mathbf{T}(\mathbf{A}))_U$ such that $[\mathbb{I}] = \mathbb{I}$. For the forest u, supposing $[u]$ be the class of all different Hall forests associated to u in $\mathbf{H}(\mathbf{F}(\mathbf{A}))_U$. So if $u = t_1 t_2 ... t_n$, then

$[u] = [t_1]...[t_n]$. On the basis of the structures of the mentioned Hopf algebras and the given commutative diagrams (6.1) and (6.2) in [44], we introduce new Hopf algebra homomorphisms.

The ladder trees l_n are divided powers with respect to the renormalization coproduct. So by sending the divided powers z_n to l_n, a new Hopf algebra homomorphism β_1 from $NSYM$ to H_F can be introduced.

Applying corollary 6.2.27 gives us a new format from H_U based on the labeled Hopf algebra $H_{CK}(A)$ (i.e. definition 6.2.22). For each forest $u = t_1...t_n$ of labeled planer rooted trees in H_F, define

$$\beta_2(u) := \sum_{v \in [u]} \pi(v). \qquad (6.2.4)$$

Let us show that the map β_2 is a Hopf algebra homomorphism. If we consider the renormalization coproduct with respect to admissible cuts, then for each rooted tree t, its coproduct can be rewritten by the formula

$$\Delta(t) = \mathbb{I} \otimes t + t \otimes \mathbb{I} + \sum_c R_c(t) \otimes P_c(t) \qquad (6.2.5)$$

such that the sum is over all proper admissible cuts of t, $R_c(t)$ is a rooted tree with the origin root of t (i.e. that part of the tree which remains connected with the root after applying the cut c) and $P_c(t)$ is a forest of rooted trees [13, 28, 29]. We know that H_{CK} and H_F have the same coproduct and further, they are connected and graded. On the other hand, one can show that

$$[\Delta(t)] = [\mathbb{I}][t] + [t][\mathbb{I}] + \sum_c [R_c(t)][P_c(t)]. \qquad (6.2.6)$$

Since π is a Hopf algebra morphism, we have

$$\beta_2(t_1...t_n) = \sum_{s_1 \in [t_1],...,s_n \in [t_n]} \pi(s_1...s_n) = \sum_{s_1 \in [t_1]} \pi(s_1)... \sum_{s_n \in [t_n]} \pi(s_n) = \beta_2(t_1)...\beta_2(t_n), \qquad (6.2.7)$$

$$\Delta(\beta_2(t)) = \Delta(\sum_{s \in [t]} \pi(s)) = \sum_{s \in [t]} \Delta\pi(s) = \sum_{\Delta(s) \in [\Delta(t)]} \pi(\Delta(s)) = \beta_2(\Delta(t)). \qquad (6.2.8)$$

Applying the surjective morphism π enables us to show that labeled ladder trees are representing divided powers in H_U. In addition, it is easy to see that $m_{(p_1,...,p_k)} = \sum_{i_1,...,i_k} x_{i_1}^{p_1} x_{i_2}^{p_2}...x_{i_k}^{p_k}$ and elementary symmetric functions $m_{(1,...,1)}$ are divided powers. Collecting these information allows us to define a new Hopf algebra morphism β_4 that sending each $m_{(1,...,1)}$ to $\sum_{v \in [l_n]} \pi(v)$.

About β_3, it is the abelianization homomorphism that sends each z_n to the elementary symmetric function of degree n.

Let us summarize the above details in the following diagram:

$$\begin{array}{ccc} z_n & \to & l_n \\ \downarrow & & \downarrow \\ m\underbrace{(1,...,1)}_{n} & \to & \sum_{v \in [l_n]} \pi(v) \end{array} \qquad (6.2.9)$$

It is possible to formulate a dual version from the above diagram. Using the theorem 6.1.1 and the corollary 6.1.9 determines the shuffle type Hopf algebra structure on $U(L_U)$. So the morphism β_3^* is the inclusion map and the morphism β_1^* is also defined by Hoffman in [44] such that for each planar rooted tree $t_J := B^+(l_{j_1},...,l_{j_k})$ (where $J = (j_1,...,j_k)$), $\beta_1^*(t_J) := M_J$.

85

For the given balanced bracket representations $c = c_1...c_n$, $c' = c'_1...c'_n$, the product $c.c'$ in $H_\mathbf{P}$ is given by shuffling the symbols $c_1, ..., c_n$ into c' and the coproduct structure is given by

$$\Delta(c) = \sum_{i=0}^{n} c_1...c_i \otimes c_{i+1}...c_n. \tag{6.2.10}$$

It is shown that the Foissy Hopf algebra H_F is self-dual and it is isomorphic to $H_\mathbf{P}$.

For each n, the set $\{t \in \beta_2^{-1}(w), \ deg(w) = n\}$ has only one element l_n. Applying this note allows us to produce the other two morphisms with respect to Hoffman's morphisms (given in [44]). For each generator e_{-n} of degree n,

$$\beta_2^\star(e_{-n}) := l_n, \quad \beta_4^\star(e_{-n}) := m_{\underbrace{(1, ..., 1)}_{n}}. \tag{6.2.11}$$

Morphisms β_2, $\beta_2^\star$ produces a correspondence between the generators of the Lie algebra $U(L_\mathbb{U})$ and Lyndon words related to H_U which means that

$$e_{-n} \leftrightarrow \sum f_{k_1}...f_{k_n}. \tag{6.2.12}$$

As the result, one can say that e_{-n}s are divided powers that map to divided powers in $H_\mathbf{P}$ and SYM, respectively. So $\beta_2^\star$, $\beta_4^\star$ are Hopf algebra morphisms.

Let us summarize the above details in the following diagram:

$$\begin{array}{ccc} m_{\underbrace{(1, ..., 1)}_{n}} & \leftarrow & e_{-n} \\ \downarrow & & \downarrow \\ M_{(1,...,1)} & \leftarrow & l_n \end{array} \tag{6.2.13}$$

□

It is shown that the Foissy Hopf algebra H_F is isomorphic to the Hopf algebra $\mathbb{C}[Y_\infty]$ of planar binary trees ([47, 74]) and the photon Hopf algebra H^γ ([7, 34]). Now if we apply the above result 6.2.29, it will be possible to obtain Hopf algebraic homomorphisms between $\mathbb{C}[Y_\infty]$, H^γ and the universal Hopf algebra of renormalization. These homomorphisms allow us to consider the structure of these Hopf algebra in the context of Hall rooted trees and Lyndon words.

Now we want to lift the Zhao's homomorphism given in the definition 2.2.10 and its dual to the level of the universal Hopf algebra of renormalization. It is shown that π is a surjective homomorphism from $H_{CK}(A)$ to $H_\mathbb{U}$. On the other hand, Z^* provides a unique surjective map from H_{CK} to $QSYM$ with the lemma 2.2.11. For a word w with length n in $H_\mathbb{U}$, there exists a labeled ladder tree l_n^w of degree n in $H_{CK}(A)$ such that $\pi(l_n^w) = w$.

Definition 6.2.30. *Define a new map* $Z_u : H_\mathbb{U} \longrightarrow QSYM$ *such that for each element* $w \in H_\mathbb{U}$, *it is given by* $Z_u(w) := Z^*(l_n^w)$.

It can be seen that Z_u is a homomorphism of Hopf algebras and it is unique with respect to the lemma 2.2.11. Applying theorem 2.2.6 and corollary 6.2.29, one can define homomorphisms

$$\theta_1 : NSYM \longrightarrow H_\mathbb{U}, \quad \theta_1 := \beta_4 \circ \beta_3 = \beta_2 \circ \beta_1, \tag{6.2.14}$$

$$\theta_2 : H_{GL} \longrightarrow H_\mathbb{U}, \quad \theta_2 := \beta_4 \circ \alpha_4^\star = \beta_2 \circ \alpha_2^\star. \tag{6.2.15}$$

Lemma 6.2.31. *The surjective morphism π induces a new homomorphism Ξ from H_{CK} to H_U such that for each unlabeled forest u in H_{CK}, it is defined by*

$$\Xi(u) := \sum_{v \in [u]} \pi(v).$$

Corollary 6.2.32. *Rooted tree reformulation of H_U determines the following commutative diagram.*

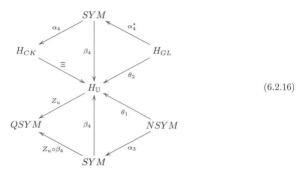

(6.2.16)

It is also possible to consider the dual version of the above diagram.

Lemma 6.2.33. *Dual of the maps Z_u and Ξ can be give by*

$$Z_u^\star : NSYM \longrightarrow U(L_U), \ Z_u^\star(z_n) := e_{-n},$$

$$\Xi^\star : U(L_U) \longrightarrow H_{GL}, \ \Xi^\star(e_{-n}) := l_n.$$

Corollary 6.2.34. *The dual version of the diagram (6.2.16) is given by*

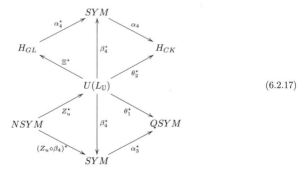

(6.2.17)

Based on the graded dual relation, there is another procedure to obtain other connections between H_U and combinatorial Hopf algebras.

Lemma 6.2.35. *Let H_1 and H_2 be graded connected locally finite Hopf algebras which admit inner products $(.,.)_1$ and $(.,.)_2$, respectively. If they are dual to each other, then there is a linear map $\lambda : H_1 \longrightarrow H_2$ such that*

(i) λ preserves degree,
(ii) For each $h_1, h_2 \in H_1 : (h_1, h_2)_1 = (\lambda(h_1), \lambda(h_2))_2$,
(iii) For each $h_1, h_2, h_3 \in H_1$:

$$(h_1 h_2, h_3)_1 = (\lambda(h_1) \otimes \lambda(h_2), \Delta_2(\lambda(h_3)))_2,$$

$$(h_1 \otimes h_2, \Delta_1(h_3))_1 = (\lambda(h_1)\lambda(h_2), \lambda(h_3))_2.$$

[44]

Remark 6.2.36. *Lemma 6.2.35 determines an isomorphism* $\tau : H_2 \longrightarrow H_1^*$ *such that for each* $h_1 \in H_1$ *and* $h_2 \in H_2$, *it is defined by*

$$< \tau(h_2), h_1 >:= (h_2, \lambda(h_1))_2.$$

Since $H_{CK}(A)$ and $H_{GL}(A)$ are graded dual to each other, therefore by 6.2.35 one can find a linear map $\lambda : H_{CK}(A) \longrightarrow H_{GL}(A)$ with the mentioned properties such that it defines an isomorphism τ_1 from $H_{GL}(A)$ to $H_{CK}(A)^*$ given by

$$< \tau_1(t), s >:= (t, \lambda(s)) \qquad (6.2.18)$$

where for rooted trees t_1, t_2, if $t_1 = t_2$ then $(t_1, t_2) = |sym(t_1)|$ and otherwise it will be 0. For each word w with length n in $H_\mathbb{U}$, there is a labeled ladder tree l_n^w in $H_{CK}(A)$ such that $\pi(l_n^w) = w$.

Proposition 6.2.37. *There is a homomorphism of Hopf algebras* $F : H_\mathbb{U} \longrightarrow H_{GL}(A)$ *given by*

$$F(w) := \tau_1^{-1}((l_n^w)^\star).$$

We know that $H_\mathbb{U}$ and $U(L_\mathbb{U})$ are graded dual to each other. And therefore lemma 6.2.35 introduces a linear map $\theta : H_\mathbb{U} \longrightarrow U(L_\mathbb{U})$ with the mentioned properties. This morphism determines an isomorphism τ_2 from $U(L_\mathbb{U})$ to $H_\mathbb{U}^*$ such that for each element $x \in U(L_\mathbb{U})$ and word $w \in H_\mathbb{U}$, we have

$$< \tau_2(x), w >:= (x, \theta(w)) \qquad (6.2.19)$$

where $(.,.)$ is the natural pairing on $U(L_\mathbb{U})$.

Proposition 6.2.38. *For a labeled forest u with the corresponding element $\pi(u)$ in $H_\mathbb{U}$, based on the natural pairing (given in theorem 6.1.1), the dual of F is clarified by*

$$F^\star : H_{CK}(A) \longrightarrow U(L_\mathbb{U}),$$

$$F^\star(u) := \tau_2^{-1}((\pi(u))^\star).$$

It should be remarked that one can extend this rooted tree reformulation from $H_\mathbb{U}$ to its related affine group scheme $\mathbb{U}$ where it is done by the theory of construction of a group from an operad [89]. Let P be an augmented set operad and $\mathbb{K}A_P = \bigoplus_n \mathbb{K}(P_n)_{\mathsf{S}_n}$ be the direct sum of its related coinvariant spaces with the completion $\widehat{\mathbb{K}A_P} = \prod_n \mathbb{K}(P_n)_{\mathsf{S}_n}$.

Lemma 6.2.39. *There is an associative monoid structure on* $\widehat{\mathbb{K}A_P}$. *[17, 92]*

Let $\mathcal{G}_P$ be the set of all elements of $\widehat{\mathbb{K}A_P}$ whose first component is the unit **1**. It is a subgroup of the set of invertible elements. One can generalize this notion to a categorical level.

Theorem 6.2.7. *There is a functor from the category of augmented operads to the category of groups.* [17, 92]

Naturally, one can expect a Hopf algebra based on subgroup $\mathcal{G}_P$.

Theorem 6.2.8. *For a given operad P, we have a commutative Hopf algebra structure on $\mathbb{K}[\mathcal{G}_P]$ given by the set of coinvariants of the operad.* [17, 92]

Proof. It is a free commutative algebra of functions on $\mathcal{G}_P$ generated by the set $(g_\alpha)_{\alpha \in A_P}$. Each $f \in \mathcal{G}_P$ can be represented by a formal sum $f = \sum_{\alpha \in A_P} g_\alpha(f)\alpha$ such that $g_1 = 1$. □

This Hopf algebra has a close relation with the incidence Hopf algebra which means that we can define a surjective Hopf algebra homomorphism

$$\eta : g_\alpha \longmapsto \frac{F_{[\alpha]}}{|Aut(\alpha)|} \tag{6.2.20}$$

from the Hopf algebra $\mathbb{K}[\mathcal{G}_P]$ to the incidence Hopf algebra H_P. Further, based on this morphism it is observed that at the level of groups, the Lie group $G_P(\mathbb{K})$ of H_P is a subgroup of the group $\mathcal{G}_P$. [17, 100]

Corollary 6.2.40. *(i) There is a surjective morphism from the Hopf algebra $\mathbb{K}[\mathcal{G}_{NAP}]$ to the Connes-Kreimer Hopf algebra H_{CK} of rooted trees.*
(ii) The complex Lie group $G(\mathbb{C})$ of H_{CK} is a subgroup of $\mathcal{G}_{NAP}$.

Proof. Referring to theorem 2.2.9 and (6.2.20) shows that the map $\rho \circ \eta : \mathbb{K}[\mathcal{G}_{NAP}] \longrightarrow H_{CK}$ is a surjective morphism and in terms of groups we will get the second claim. □

It is discussed that the Lie group $\mathbb{U}$ of $H_\mathbb{U}$ determines universal counterterms. So $\mathbb{U}$ plays a practical role and it will be very useful to obtain a rooted tree type representation from elements of this Lie group. For the operad NAP, $\mathcal{G}_{NAP}$ is a group of formal power series indexed by the set of unlabeled rooted trees. Applying corollary 6.2.40 allows us to reformulate elements of $\mathbb{U}$ based on these formal power series.

Proposition 6.2.41. *(i) Group $\mathbb{U}(\mathbb{C})$ is a subgroup of $\mathcal{G}_{NAP}$. It identifies elements of $\mathbb{U}(\mathbb{C})$ and therefore each of its elements as a formal power series indexed with Hall rooted trees.*
(ii) Formal series $K = \sum_t g_t(K)t$ in $\mathcal{G}_{NAP}$ belongs to $\mathbb{U}(\mathbb{C})$ if and only if
- t be a Hall tree in $\mathbf{H}(\mathbf{T}(\mathbf{A}))_\mathbb{U}$,
- If $t = B^+_{f_n}(u)$ for some $u = t_1...t_k$ such that $t_1,...,t_k \in \mathbf{H}(\mathbf{T}(\mathbf{A}))_\mathbb{U}$ and $f_n \in A$, then

$$g_t(K) = \prod_{i=1}^{k} g_{B^+_{f_n}(t_i)}(K).$$

Proof. (i) First define a surjective homomorphism from $\mathbb{C}[\mathcal{G}_{NAP}](A)$ to $H_\mathbb{U}$. For this aim, extend the homomorphism (6.2.20) to the level of decorated Hopf algebras and then apply corollaries 6.2.27 and 6.2.40. The resulted homomorphism determines an injective morphism from $\mathbb{U}(\mathbb{C})$ to $\mathcal{G}_{NAP}$ which means that in terms of groups, $\mathbb{U}(\mathbb{C})$ is a subgroup of $\mathcal{G}_{NAP}$.
(ii) Applying [17] shows that each element of $\mathcal{G}_{NAP}$ is in the subgroup $G_{NAP}(\mathbb{C})$ iff for each tree $t = B^+(t_1,...,t_k)$, we have the following condition $|sym(t)|g_t(K) = \prod_{i=1}^{k} |sym(B^+(t_i))|g_{B^+(t_i)}(K)$. Now it is enough to apply lemma 2.2.28, theorem 2.2.9, corollaries 6.2.27, 6.2.40 and the relation (6.2.20), and since Hall trees have no symmetries, the proof is completed. □

There is another interpretation from the universal affine group scheme $\mathbb{U}^* := \mathbb{U} \rtimes \mathbb{G}_m$ as a motivic Galois group for a category of mixed Tate motives. Letting $\mathbb{Q}(\zeta_N)$ be the cyclotomic field of level N and $\mathcal{O}$ be its ring of integers. In one hand, for $N = 3$ or 4, the motivic Galois group of the category $\mathcal{MT}_{mix}(S_N)$ of mixed Tate motives (where $S_N = Spec(\mathcal{O}[\frac{1}{N}])$) is non-canonically of the form $\mathbb{U}^*$. On the other hand, for the scheme $S_N = Spec(\mathbb{Z}[i][\frac{1}{2}])$ of N−cyclotomic integers, there is a non-canonical isomorphism between $\mathbb{U}^*$ and the motivic Galois group $G_{\mathcal{M}_T}(\mathbb{Z}[i][\frac{1}{2}])$ [21, 22]. Applying corollaries 6.2.40 and 6.2.41 allows us to map (at the level of groups) the formulated Hall tree interpretation to the level of these motivic Galois groups.

Since the universal Hopf algebra of renormalization is a resulted object from the Connes-Kreimer perturbative renormalization theory, so the study of this Hopf algebra at the level of its Lie algebra is also important. We know that this Lie algebra can be identified by the well-know Milnor-Moore theorem 2.1.3. Now we want to provide a new picture from elements of this Lie algebra at the level of Hall rooted trees. Actually, this new interpretation help us to characterize PBW basis and Hall basis with respect to the Hall set $\mathbf{H}(\mathbf{T}(\mathbf{A}))_\mathbb{U}$.

Definition 6.2.42. *Let $H(A)$ be a Hall set and $A^* := \{a^* : a \in A\}$. For each Hall word w, its associated Hall polynomial p_w in the free Lie algebra $\mathcal{L}(A^*)$ is introduced by*
- *If $f_{k_j} \in A$, then $p_{f_{k_j}} = f^*_{k_j}$,*
- *If w be a Hall word of length ≥ 2 such that its corresponding Hall tree t_w has the standard decomposition (t_{w_1}, t_{w_2}), then $p_w = [p_{w_1}, p_{w_2}]$.*

Lemma 6.2.43. *(i) With induction one can show that each p_w is a homogeneous Lie polynomial of degree equal to the length of w and also, it has the same partial degree with respect to each letter as w.*
(ii) For a given Hall set H, Hall polynomials form a basis for the free Lie algebra (viewed as a vector space) and their decreasing products $p_{f_{k_1}}...p_{f_{k_n}}$ such that $f_{k_i} \in H$, $f_{k_1} > f_{k_2} > ... > f_{k_n}$, form a basis for the free associative algebra (viewed as a vector space) [88].

We found a Hall set $\mathbf{H}(\mathbf{T}(\mathbf{A}))_\mathbb{U}$ corresponding to $H_\mathbb{U}$ such that $A = \{f_n\}_{n \in \mathbb{N}_{>0}}$ and for each f_n its associated Hall polynomial is given by

$$p_{f_n} = e_{-n}. \qquad (6.2.21)$$

Corollary 6.2.44. *Hall polynomials associated to the Hall set $\mathbf{H}(\mathbf{T}(\mathbf{A}))_\mathbb{U}$ determines a basis (as a vector space) for the Lie algebra $L_\mathbb{U}$. Decreasing products of Hall polynomials with respect to the Hall set $\mathbf{H}(\mathbf{T}(\mathbf{A}))_\mathbb{U}$ determines a basis (as a vector space) for the free algebra $H_\mathbb{U}$.*

Therefore in summary, based on the basic set operad NAP on rooted trees, one can provide a complete reconstruction from the specific Hopf algebra $H_\mathbb{U}$ in the algebraic perturbative renormalization, its related infinite dimensional complex Lie group and also Lie algebra (with notice to the Milnor-Moore theorem). Now it can be useful to apply this induced version in the Connes-Marcolli's universal treatment such that consequently, new reformulations from physical information such as counterterms can be expected. This project is actually performed by universal singular frame.

6.3 Universal counterterms

The appearance of the Connes-Marcolli categorification of renormalization theory depends directly upon the Riemann-Hilbert problem such that as the result one can search the application of the theory of motives in describing divergences. Neutral Tannakian formalism of the category of flat equi-singular vector bundles could determine the special loop $\gamma_\mathbb{U}$ with values in the universal affine group scheme $\mathbb{U}(\mathbb{C})$. Since now we have enough knowledge about the combinatorics of the universal Hopf algebra of renormalization, so it does make sense to search for a combinatorial nature inside of the universal singular frame. The explanation of this important frame (free of any divergences) based on Hall basis and PBW basis is the main purpose of this section such that universality of $\gamma_\mathbb{U}$ allows us to modify this new Hall type representation to counterterms of any renormalizable theory.

Starting with the application of words in geometry. Let M be a smooth manifold, $C(\mathbb{R}^+)$ the ring of real valued piecewise continuous functions on $\mathbb{R}^+$ and $\{X_a\}_{a \in A}$ a family of smooth vector fields on M. Suppose $\mathcal{A}$ be the algebra over $C(\mathbb{R}^+)$ of linear operators on $C^\infty(M)$ generated by the vector fields X_a ($a \in A$). For a family $\{g_a\}_{a \in A}$ of elements in $C(\mathbb{R}^+)$, set

$$X(x) = \sum_{a \in A} g_a(x) X_a. \tag{6.3.1}$$

Lemma 6.3.1. *The formal series (6.3.1) can be expanded as a series of linear operators in $\mathcal{A}$ of the form $\sum_w g_w X_w$ such that*
- *$w = a_1...a_m$ is a word on A,*
- *$X_1 = Id$ (identity operator), $X_w = X_{a_1}...X_{a_m}$,*
- *$g_w = \int_{a_m} \cdots \int_{a_1} 1_{C(\mathbb{R}^+)}$ where each $\int_{a_i} : C(\mathbb{R}^+) \longrightarrow C(\mathbb{R}^+)$, ($1 \leq i \leq m$) is a linear endomorphism defined by*

$$\{\int_{a_i} g\}(x) := \int_0^x g(s) g_{a_i}(s) ds.$$

This technique can be generalized to any arbitrary algebra.

Theorem 6.3.1. *For a given associative algebra $\mathcal{A}$ over the field $\mathbb{K}$ of characteristic zero generated by the elements $\{E_a\}_{a \in A}$, all elements in $\mathcal{A}$ are characterized by formal series $\sum_w \mu_w E_w$ such that $\mu_w \in \mathbb{K}$. [77]*

Remark 6.3.2. *If $\mathcal{A}$ is a free algebra, then this formal series representation is unique.*

Definition 6.3.3. *Suppose $\mathfrak{a}$ be the unital Lie algebra generated by the set $\{E_a\}_{a \in A}$. For a given Hall set $\mathbf{H}(\mathbf{T}(\mathbf{A}))$ of labeled rooted trees with the corresponding Hall forest $\mathbf{H}(\mathbf{F}(\mathbf{A}))$, one can assign elements $E(u)$ such that*
- *$E(\mathbb{I}) = \mathfrak{e}$ (The unit of $\mathfrak{a}$),*
- *For each Hall tree t with the standard decomposition $(t^1, t^2) \in \mathbf{H}(\mathbf{T}(\mathbf{A})) \times \mathbf{H}(\mathbf{T}(\mathbf{A}))$,*

$$E(t) = [E(t^2), E(t^1)] = E(t^2)E(t^1) - E(t^1)E(t^2),$$

- *For each $u \in \mathbf{H}(\mathbf{F}(\mathbf{A})) - \mathbf{H}(\mathbf{T}(\mathbf{A}))$ with the standard decomposition $(u^1, u^2) \in \mathbf{H}(\mathbf{F}(\mathbf{A})) \times \mathbf{H}(\mathbf{T}(\mathbf{A}))$,*

$$E(u) = E(u^2)E(u^1).$$

Lemma 6.3.4. *(i) The Lie algebra $\mathfrak{a}$ is spanned by $\{E(t) : t \in \mathbf{H}(\mathbf{T}(\mathbf{A}))\}$ (as an ordered basis). It is called Hall basis.*
(ii) $\mathcal{A}$ is spanned by $\{E(u) : u \in \mathbf{H}(\mathbf{F}(\mathbf{A}))\}$. It is called PBW basis. [77]

It is the place to give a new understanding from the universal singular frame.

Proposition 6.3.5. *For the locally finite total order set $\{f_n\}_{n\in\mathbb{N}}$, the universal singular frame can be rewritten by*

$$\gamma_U(-z,v) = \sum_{n\geq 0, k_j > 0} \alpha^U_{f_{k_1} f_{k_2} \cdots f_{k_n}} p_{f_{k_1}} \cdots p_{f_{k_n}} v^{\sum k_j} z^{-n}$$

such that $p_{f_{k_j}}$s are Hall polynomials.

Proof. From theorem 5.2.4, we know that

$$\gamma_U(z,v) = Te^{-\frac{1}{z}\int_0^v u^Y(e)\frac{du}{u}}.$$

After the application of the time ordered exponential, one can get

$$\gamma_U(-z,v) = \sum_{n\geq 0, k_j > 0} \frac{e_{-k_1}\cdots e_{-k_n}}{k_1(k_1+k_2)\cdots(k_1+\cdots+k_n)} v^{\sum k_j} z^{-n}$$

such that in this expansion the coefficient of $e_{-k_1}\cdots e_{-k_n}$ is calculated by the iterated integral

$$\int_{0\leq s_1 \leq \cdots \leq s_n \leq 1} s_1^{k_1-1}\cdots s_n^{k_n-1} ds_1 \cdots ds_n.$$

On the other hand, theorem 6.2.3 determines that Hopf algebra H_U is a free polynomial algebra on Lyndon words on the set $\{f_n\}_{n\in\mathbb{N}_{>0}}$. Consider formal series

$$E := f_{k_1} + x f_{k_2} + x^2 f_{k_3} + \cdots$$

where

$$\mu_{k_j}(x) = x^{k_j - 1}.$$

By the lemma 6.3.1, for the variables $0 \leq s_1 \leq \cdots \leq s_n \leq 1$, we have

$$\{\int_{k_j} 1\}(s_j) = \int_0^{s_j} x^{k_j-1} dx.$$

For each word $f_{k_1} f_{k_2} \cdots f_{k_n}$, we can define the following well-defined iterated integral

$$\alpha^U_{f_{k_1} f_{k_2} \cdots f_{k_n}} := \int_{k_n} \cdots \int_{k_1} 1.$$

It is easy to see that the above integral is agree with the iterated integral associated to the coefficient of the term $e_{-k_1}\cdots e_{-k_n}$. This completes the proof. □

In addition, one can determine uniquely a real valued map on the set $\mathbf{F}(\mathbf{A})$.

Definition 6.3.6. Let $A = \{f_n\}_{n \in \mathbb{N}}$ be the locally finite total order set corresponding to the universal Hopf algebra of renormalization. For the given map in the above proof that associates to each word $w = f_{k_1} f_{k_2} ... f_{k_n}$ a real value α_w^{U} and based on the definition 6.2.10, we introduce a new map α^{U} on $\mathbf{F}(\mathbf{A})$ such that
- $\mathbb{I} \longmapsto \alpha^{\mathrm{U}}(\mathbb{I}) = 1$,
- For each non-empty labeled forest u in $\mathbf{F}(\mathbf{A})$,

$$\alpha^{\mathrm{U}}(u) = \sum_{>_{u(A)}} \alpha_{w(>_{u(A)})}^{\mathrm{U}}.$$

Lemma 6.3.7. For given labeled rooted trees $t_1, ..., t_m \in \mathbf{T}(\mathbf{A})$ and $f_{k_j} \in A$,

$$\alpha^{\mathrm{U}}(t_1...t_m) = \alpha^{\mathrm{U}}(t_1)...\alpha^{\mathrm{U}}(t_m), \quad \alpha^{\mathrm{U}}(B_{f_{k_j}}^+(t_1...t_m)) = \int_{k_j} \alpha^{\mathrm{U}}(t_1)...\alpha^{\mathrm{U}}(t_m).$$

The map α^{U} (determined by the universal singular frame), with the above properties, can be uniquely identified.

Lemma 6.3.8. For the given map in definition 6.3.6 together with the mentioned properties in lemma 6.3.7, there exists a real valued map β^{U} on $\mathbf{F}(\mathbf{A})$ given by

$$\alpha^{\mathrm{U}} = exp\ \beta^{\mathrm{U}}$$

such that for each $u \in \mathbf{F}(\mathbf{A}) - \mathbf{T}(\mathbf{A})$, $\beta^{\mathrm{U}}(u) = 0$.

Proof. (A sketch of the proof.) For a given map $\alpha : \mathbf{F}(\mathbf{A}) \longrightarrow \mathbb{K}$,
If $\alpha(\mathbb{I}) = 0$, then the exponential map is defined by

$$exp\ \alpha(\mathbb{I}) = 1,$$

and for each $u \in \mathbf{F}(\mathbf{A}) - \{\mathbb{I}\}$,

$$exp\ \alpha(u) = \sum_{k=1}^{|u|} \frac{1}{k!} \alpha^k(u).$$

And if $\alpha(\mathbb{I}) = 1$, then the logarithm map is defined by

$$log\ \alpha(\mathbb{I}) = 0,$$

and for each $u \in \mathbf{F}(\mathbf{A}) - \{\mathbb{I}\}$,

$$log\ \alpha(u) = \sum_{k=1}^{|u|} \frac{(-1)^{k+1}}{k} (\alpha - \epsilon)^k(u).$$

such that $\epsilon(\mathbb{I}) = 1$ and for $u \in \mathbf{F}(\mathbf{A}) - \{\mathbb{I}\}$, $\epsilon(u) = 0$. With the help of proposition 7 in [77], the proof is completed. □

Lemma 6.3.9. (i) The set of all Hall polynomials $\{E(t) : t \in \mathbf{H}(\mathbf{T}(\mathbf{A}))_{\mathrm{U}}\}$ is the Hall basis for L_{U},
(ii) The set of decreasing products of Hall polynomials $\{E(u) : u \in \mathbf{H}(\mathbf{F}(\mathbf{A}))_{\mathrm{U}}\}$ is the PBW basis

for H_U.
(iii) Applying the mentioned Hall basis and PBW basis provides the relation

$$\sum_{k_j>0, n\geq 0} \alpha^U_{f_{k_1} f_{k_2} \cdots f_{k_n}} f_{k_1} f_{k_2} \cdots f_{k_n} = exp \left(\sum_{t\in \mathbf{H}(\mathbf{T}(\mathbf{A}))_U} \beta^U(t) E(t) \right)$$

such that

Proof. One can show that for a given map $\alpha : \mathbf{F}(\mathbf{A}) \longrightarrow \mathbb{K}$, if $\alpha(\mathbb{I}) = 0$, then

$$exp \left(\sum_w \alpha_w E_w \right) = \sum_{u\in \mathbf{H}(\mathbf{F}(\mathbf{A}))} exp\ \alpha(u) E(u),$$

and if $\alpha(\mathbb{I}) = 1$, then

$$log \left(\sum_w \alpha_w E_w \right) = \sum_{u\in \mathbf{H}(\mathbf{F}(\mathbf{A}))} log\ \alpha(u) E(u).$$

Now by lemma 6.3.7 and with help of the continuous BCH formula (44) in [77], for each word $w = f_{k_1} f_{k_2} \cdots f_{k_n}$ on the set A, one can have

$$exp \left(\sum_{t\in \mathbf{H}(\mathbf{T}(\mathbf{A}))_U} \beta^U(t) E(t) \right) = \sum_w \alpha^U_w w.$$

Hall basis and PBW basis depended on H_U can be determined by corollary 6.2.44 and lemma 6.3.4. □

With attention to lemma 6.3.4, definition 6.3.6 and proposition 6.3.5, a Hall representation from γ_U is obtained.

Definition 6.3.10. *The formal series $\sum_{t\in \mathbf{H}(\mathbf{T}(\mathbf{A}))_U} \beta^U(t) E(t)$ becomes a Hall polynomial representation for the universal singular frame.*

Corollary 6.2.40 and relation (6.2.20) which give us a surjective morphism from $\mathbb{C}[\mathcal{G}_{NAP}](A)$ to the Hopf algebra H_U provide another procedure to rewrite universal singular frame with respect to rooted trees. In addition applying corollary 6.2.41 shows that the infinite dimensional complex Lie group $\mathbb{U}(\mathbb{C})$ is indeed a subgroup of $\mathcal{G}_{NAP}$. We know that γ_U is a loop with values in $\mathbb{U}(\mathbb{C})$. So for each fixed values z, v, $\gamma_U(z, v)$ is a formal power series of Hall rooted trees which satisfies the mentioned conditions in the corollary 6.2.40.

Now let us come to conclusion. In [28, 29] one can find rooted tree type reformulations of components of the Birkhoff decomposition of a dimensionally regularized Feynman rules character based on Baker-Campbell-Hausdorff (BCH) series and Bogoliubov character such that as the result, scattering formulaes for the renormalization group and the β-function can be investigated. Now here we can obtain a new Hall rooted tree scattering type formula for counterterms based on the given combinatorial version of H_U.

Corollary 6.3.11. *The Hall polynomial representation of the universal singular frame can be mapped to counterterms.*

Proof. Applying the graded representation $\xi_{\gamma_\mu} : \mathbb{U} \longrightarrow G$ identified by theorem 1.106 in [22] can help us to map the universal singular frame γ_U to the negative part $\gamma_-(z)$ of the Birkhoff decomposition of a loop $\gamma_\mu(z)$ in $Loop(G(\mathbb{C}), \mu)$. On the other hand, the minus part which is independent of μ (such that μ is the mass parameter) does determine counterterms. Now map Hall tree type formula obtained by propositions 6.3.5 and 6.3.9 underlying ξ_{γ_μ}. □

Since Hall set $\mathbf{H}(\mathbf{T}(\mathbf{A}))_U$ and its related Hall polynomials determine Hall basis and PBW basis, one can expect to reproduce the BCH type representations of physical information (given in [28, 29]) with respect to this Hall tree approach. This possibility reports another reason for the importance of H_U in the study of renormalizable theories.

Corollary 6.3.12. *On the basis of formal sums of Hall trees (Hall forests) and Hall polynomials associated to the universal Hopf algebra of renormalization, universal counterterms can be represented by these rooted tree type formal sums. This introduces new reformulations from divergences (counterterms) of each arbitrary renormalizable theory such that they can be applied to redefine the renormalization group and its infinitesimal generator on the basis of Hall basis and PBW basis.*

Chapter 7

Combinatorial Dyson-Schwinger Equations and Connes-Marcolli Universal Treatment

The study of Dyson-Schwinger equations (DSEs) could help people as an influenced method of describing unknown nonperturbative circumstances and in fact, it has central role in the development of modern physics. These equations enable us to find an effective conceptional explanation from nonperturbative theory which is the main complicate part of quantum field theory. It seems that a nonperturbative theory can be discovered with solving its corresponding DSE. The significance of this approach for the general identification of quantum field theory is on the increase. [73, 91]

In modern physics people concentrate on an analytic approach to this type of practical equations such that it contains a class of equations related to ill-defined iterated Feynman integrals. More interestingly, Kreimer introduces a new combinatorial reinterpretation from this class of equations such that in his strategy, he systematically applies the combinatorics of renormalization together with Hochschild cohomology theory to obtain perturbative expansions of Hochschild one cocycles depended upon Hopf algebra of Feynman diagrams. This process justifies combinatorial Dyson-Schwinger equations (DSEs) and so it will be considerable to derive nonperturbative theory from the capsulate renormalization Hopf algebra. It should be mentioned that in principle, this formalism is strongly connected with the analytic formulation. Because Kreimer could regain analytic DSEs from this combinatorial version by using one particular measure which relates Feynman rules characters (in the Hopf algebraic language) with their corresponding standard forms. [8, 9, 54, 56, 57, 59, 60, 61]

In short, on the one hand, DSEs report nonperturbative phenomena and on the other hand, they can be formalized on the basis of infinite formal perturbative expansions of one cocycles. So it means that the formalization of quantum field theory underlying Hopf algebra of renormalization leads us widely to clarify a more favorable explanation from nonperturbative quantum field theory.

In this chapter at first we consider Kreimer's programme in the study of DSEs based on Hochschild cohomology theory on various Hopf algebras of rooted trees. Then after redefining one cocycles at the Lie algebra level, we consider DSEs at the level of the universal Hopf algebra of renormalization (with attention to its described combinatorics). In the next step, we want to show the importance of H_U in the theory of DSEs and this work will be done based on the factorization principal of Feynman diagrams into primitive components. As the result, we will extend reasonably the universality of H_U (i.e. independency of all renormalizable physical theories) to nonperturbative theory. And finally, we show that the study of DSEs at the level of the universal Hopf algebra of renormalization hopefully leads to a new explanation from this kind of equations in a combinatorial-categorical configuration. This procedure provides one important fact that the universality of the category of equi-singular flat vector bundles can be developed to DSEs and therefore nonperturbative theory. [97, 99]

Totally, these observations mean that beside analytic and combinatorial techniques, one can expect a new categorical geometric interpretation from this family of equations.

7.1 Quantum motions in terms of the renormalization Hopf algebra

One crucial property of DSEs is their ability in describing the self-similarity nature of amplitudes in renormalizable QFTs such that it can help us to introduce a recursive combinatorial version from these equations. Starting with the physical theory Φ with the associated Hopf algebra $H_{FG} = H(\Phi)$. With applying the renormalization coproduct, one can indicate a coboundary operator which introduces a Hochschild cohomology theory with respect to the Hopf algebra H_{FG}.

Definition 7.1.1. *Define a chain complex $(C, \mathbf{b})$ with respect to the coproduct structure of H such that the set of $n-$cochains and the coboundary operator are determined with*

$$C^n := \{T : H \longrightarrow H^{\otimes n} : Linear\},$$

$$\mathbf{b}T := (id \otimes T)\Delta + \sum_{i=1}^{n}(-1)^i \Delta_i T + (-1)^{n+1} T \otimes \mathbb{I}$$

where $T \otimes \mathbb{I}$ is given by $x \longmapsto T(x) \otimes \mathbb{I}$.

Lemma 7.1.2. *(i) $\mathbf{b}^2 = 0$,*
(ii) H is a bicomodule over itself with the right coaction $(id \otimes \epsilon)\Delta$.

Proposition 7.1.3. *The cohomology of the complex $(C, \mathbf{b})$ is denoted by $HH_\epsilon^\bullet(H)$ and one can show that for $n \geq 2$, $HH_\epsilon^n(H)$ is trivial. [8]*

It should be remarked that this definition of the coboundary operator strongly connected with the Connes-Kreimer coproduct which yields the universality of the pair (H_{CK}, B^+) with respect to Hochschild cohomology theory. Indeed, $\mathbf{b}$ is defined in the sense that the grafting operator B^+ can be deduced as a Hochschild one cocycle.

$$\mathbf{b}T = 0 \iff \Delta(T) = (id \otimes T)\Delta + T \otimes \mathbb{I}. \tag{7.1.1}$$

Moreover, it is remarkable to see that these one cocycles determine generators of $HH_\epsilon^1(H)$. Thus it is necessary to collect more information about Hochschild one cocycles.

Lemma 7.1.4. *(i) Let T be a generator of $HH^1_\epsilon(H)$, then $T(\mathbb{I})$ is a primitive element of the Hopf algebra.*
(ii) There is a surjective map $HH^1_\epsilon(H) \longrightarrow Prim(H)$, $T \longmapsto T(\mathbb{I})$.
(iii) We can translate 1-cocycles to the universal enveloping algebra on the dual side and it means that for example 1-cocycle $B^+ : H \longrightarrow H_{lin}$ turns out to the dual map $(B^+)^\star : \mathcal{L} \longrightarrow U(\mathcal{L})$ where that is a 1-cocycle in the Lie algebra cohomology. [8, 63]

So it can be seen that primitive elements of the renormalization Hopf algebra and the grafting operator B^+ can introduce an important part of the generators of $HH^1_\epsilon(H)$. In fact, the sum over all primitive graphs (i.e. graphs without any divergent subgraphs which need renormalization) of a given loop order n defines a 1-cocycle B_n^+ such that every graph is generated in the range of these 1-cocycles. The importance of cocycles in the reformulation of Green functions will be observed more clear, if we understand that every relevant tree or Feynman graph is in the range of a homogeneous Hochschild 1-cocycle of degree one.

One should notice that insertion into a primitive graph commutes with the coproduct and therefore it determines a generator of $HH^1_\epsilon(H_{CK}(\Phi))$.

Lemma 7.1.5. *In a renormalizable theory Φ, for each $r \in \mathcal{R}_+$, we know that $G^r_\phi = \phi(\Gamma^r)$. There are Hochschild 1-cocycles $B^+_{k,r}$ in their expansions such that they can be formulated with building blocks (i.e. 1PI primitives graphs) of the theory and it means that*

$$B^+_{k,r} = \sum_{\gamma \in H(\Phi)^{(1)} \cap H_{lin}(\Phi), res(\gamma)=r} \frac{1}{|sym(\gamma)|} B^+_\gamma$$

where the sum is over all Hopf algebra primitives γ contributing to the amplitude r at k loops. [36, 55, 62]

In gauge theories, we may have some overlapping sub-divergences with different external structures. Therefore it is possible to make a graph Γ by inserting one graph into another but in the coproduct of Γ there may be subgraphs and cographs completely different from those which we used to make Γ. In this situation, it is impossible that each B^+_Γ (that Γ is 1PI) be a Hochschild 1-cocycle. Since there may be graphs appearing on the right hand side of the formula in (7.1.1) which do not appear on the left hand side and for solving this problem in these theories, there are identities between graphs. For example: Ward identities in QED and Slavnov-Taylor (ST) identities in QCD. These identities generate Hopf (co-)ideals in $H_{CK}(QED)$ and $H_{CK}(QCD)$, respectively where with working on the related quotient Hopf algebras, we will obtain Hochschild 1-cocyles. [60, 61, 64, 70, 102, 103, 112]

Lemma 7.1.6. *The locality of counterterms and the finiteness of renormalized Green functions are another applications of Hochschild cohomology theory in QFT. [9, 16, 57, 62]*

Another important influence of Hochschild cohomology in quantum field theory can be found in rewriting quantum equations of motion in terms of Hopf algebra primitives and elements in

$$H_{lin}(\Phi) \cap \{ker\,Aug^{(2)}/ker\,Aug^{(1)}\} \tag{7.1.2}$$

such that it can be applied to characterize DSEs.

Definition 7.1.7. *For a given renormalizable theory Φ, let H be its associated free commutative connected graded Hopf algebra and $(B_{\gamma_n}^+)_{n \in \mathbb{N}}$ be a collection of Hochschild 1-cocycles (related to the primitive 1PI graphs in H). A class of combinatorial Dyson-Schwinger equations in $H[[\alpha]]$ has the form*

$$X = \mathbb{I} + \sum_{n \geq 1} \alpha^n \omega_n B_{\gamma_n}^+(X^{n+1})$$

such that $\omega_n \in \mathbb{K}$ and α is a constant.

This family of equations provides a new source of Hopf subalgebras such that with working on rooted trees, some interesting relations between Hopf subalgebras of rooted trees (decorated by primitive 1PI Feynman graphs) and Dyson-Schwinger equations (at the combinatorial level) can be found.

Theorem 7.1.1. *Each combinatorial Dyson-Schwinger equation DSE has a unique solution (given by $c = (c_n)_{n \in \mathbb{N}}$, $c_n \in H$) such that it generates a Hopf subalgebra of H. [8, 9, 63]*

Proof. The elements c_n are determined by

$$c_0 = \mathbb{I},$$

$$c_n = \sum_{m=1}^{n} \omega_m B_{\gamma_m}^+ \big(\sum_{k_1+\ldots+k_{m+1}=n-m, k_i \geq 0} c_{k_1} \ldots c_{k_{m+1}} \big).$$

And so the unique solution is given by $X = \sum_{n \geq 0} \alpha^n c_n$. The related Hopf subalgebra structure from this unique solution is given by the following coproduct

$$\Delta(c_n) = \sum_{k=0}^{n} P_k^n \otimes c_k$$

such that $P_k^n := \sum_{l_1+\ldots+l_{k+1}=n-k} c_{l_1} \ldots c_{l_{k+1}}$ are homogeneous polynomials of degree $n - k$ in c_l ($l \geq n$). □

Remark 7.1.8. *The independency of the coproduct from the scalars ω_k determines an isomorphism between the induced Hopf subalgebras by all DSEs of this class. It means that all Hopf subalgebras which come from a fixed general class of DSEs are isomorphic.*

Lemma 7.1.9. *For a connected graded Hopf algebra H and an element $a \in H$, set*

$$val(a) := max\{n \in \mathbb{N} : a \in \bigoplus_{k \geq n} H_k\}.$$

It defines a distance on H such that for each $a, b \in H$,

$$d(a, b) := 2^{-val(a-b)}$$

where its related topology is called n-adic topology.

Lemma 7.1.10. *The completion of H with this topology is denoted by $\overline{H} = \prod_{n=0}^{\infty} H_n$ such that its elements are written in the form $\sum a_n$ where $a_n \in H_n$. One can show that $\overline{H}$ has a Hopf algebra structure originally based on H and in fact, the solution of a DSE belongs to this completion. [35, 36]*

Since Hopf algebra of Feynman diagrams of a given theory fundamentally defined by the insertion of graphs into each other (namely, pre-Lie operation $\star$), therefore one can have a new description from Hochschild one cocycles in terms of this operator. In continue we consider this notion.

Definition 7.1.11. *For the Lie algebra $\mathfrak{g}$, let $U(\mathfrak{g})$ be its universal enveloping algebra. For a given $\mathfrak{g}$-module M (with the Lie algebra morphism $\mathfrak{g} \longrightarrow End_{\mathbb{K}}(M)$), one can define a $U(\mathfrak{g})$-bimodule structure given by $M^{ad} = M$ with left and right $U(\mathfrak{g})$-actions: $X.m = X(m)$, $m.X = 0$.*

Definition 7.1.12. *Suppose $C^n_{Lie}(\mathfrak{g}, M) := Hom(\bigwedge^n \mathfrak{g}, M)$ be the set of all alternating n-linear maps $f(X_1, ..., X_n)$ on $\mathfrak{g}$ with values in M. The Chevalley-Eilenberg complex is defined by*

$$M \longrightarrow^\delta C^1(\mathfrak{g}, M) \longrightarrow^\delta C^2(\mathfrak{g}, M) \longrightarrow ...$$

such that

$$\delta(f)(X_1, ..., X_{n+1}) :=$$

$$\sum_{i<j}(-1)^{i+j} f([X_i, X_j], ..., \widehat{X_i}, ..., \widehat{X_j}, ..., X_{n+1}) + \sum_i (-1)^i X_i.f(X_1, ..., \widehat{X_i}, ..., X_{n+1}).$$

Cohomology of the chain complex $(C^n_{Lie}(\mathfrak{g}, M), \delta)$ is called Lie algebra cohomology.

Theorem 7.1.2. *There is an isomorphism between Lie algebra (co)homology and Hochschild (co)homology. In other words,*

$$H^n_{Lie}(\mathfrak{g}, M) \cong HH^n(U(\mathfrak{g}), M).$$

[51]

Now let $\mathcal{L}$ be the Lie algebra (which comes from the pre-Lie operation $\star$) of Feynman diagrams of a given renormalizable theory. Because of the importance of one cocycles just we work on the case $n = 1$ and calculate the Lie algebra cohomology $H^1_{Lie}(\mathcal{L}, U(\mathcal{L}))$. We have

$$C^1_{Lie}(\mathcal{L}, U(\mathcal{L})) = \{f : \mathcal{L} \longrightarrow U(\mathcal{L}) : Linear, Alternating\}, \tag{7.1.3}$$

$$\delta f(X_1, X_2) = -f([X_1, X_2]) - X_1 f(X_2) + X_2 f(X_1) \tag{7.1.4}$$

such that X_1, X_2 are disjoint unions of 1PI graphs. Moreover

$$\delta f = 0 \iff f([X_1, X_2]) = X_2 f(X_1) - X_1 f(X_2) \tag{7.1.5}$$

where it means that

$$f(X_1 \star X_2 - X_2 \star X_1) = f(\sum_{\Gamma_1 : 1PI} n(X_1, X_2, \Gamma_1)\Gamma_1 - \sum_{\Gamma_2 : 1PI} n(X_2, X_1, \Gamma_2)\Gamma_2)$$

$$\implies \sum_{\Gamma_1, \Gamma_2 : 1PI} (n(X_1, X_2, \Gamma_1) - n(X_2, X_1, \Gamma_2))f(\Gamma_1 - \Gamma_2) = X_2 f(X_1) - X_1 f(X_2) \tag{7.1.6}$$

such that $\Gamma_1/X_2 = X_1$, $\Gamma_2/X_1 = X_2$. It is an explicit condition for one cocycles at the Lie algebra level.

Proposition 7.1.13. *Let H be the Hopf algebra of Feynman diagrams of the theory Φ and L : $H \longrightarrow H_{lin}$ a linear map with its dual $L^* : \mathcal{L} \longrightarrow U(\mathcal{L})$. Then*

$$\Delta(L) = (id \otimes L)\Delta + L \otimes \mathbb{I} \iff$$

$$\sum_{\Gamma_1, \Gamma_2 : 1PI} (n(X_1, X_2, \Gamma_1) - n(X_2, X_1, \Gamma_2))L^*(\Gamma_1 - \Gamma_2) = X_2.L^*(X_1) - X_1.L^*(X_2)$$

where Γ_1, Γ_2 are 1PI graphs such that $\Gamma_1/X_2 = X_1$, $\Gamma_2/X_1 = X_2$.

Corollary 7.1.14. *For the Connes-Kreimer Hopf algebra of rooted trees, we have*

$$H^n_{Lie}(\mathcal{L}_{CK}, U(\mathcal{L}_{CK})) \simeq HH^n(U(\mathcal{L}_{CK}), U(\mathcal{L}_{CK})) \simeq HH^n(H_{GL}, H_{GL}).$$

Result 7.1.14 shows that when we want to identify one cocycles for the Hopf algebra H_{CK} (at the Lie algebra level), it is enough to find Hochschild one cocycles on the Hopf algebra H_{GL}.

It is possible to do above procedure for the universal Hopf algebra of renormalization to characterize its corresponding one cocycles at the Lie algebra level. It was shown that $H_U \simeq \frac{\mathbb{K}[T(A)]}{I_\pi}$ such that $A = (f_n)_{n \in \mathbb{N}}$. Therefore for each element u in $\frac{\mathbb{K}[T(A)]}{I_\pi}$, there exist $t \in \mathbf{H}(\mathbf{T}(\mathbf{A}))_U$ and $i \in I_\pi$ such that

$$u = t + i. \tag{7.1.7}$$

And furthermore there exist $t_1, t_2, z_1, z_2, s \in \mathbf{T(A)}$ such that

$$u = t + t_1 \circ z_1 + z_1 \circ t_1 - t_1 z_1 + s \circ (t_2 z_2) + z_2 \circ (t_2 s) + t_2 \circ (s z_2) - t_2 z_2 s. \tag{7.1.8}$$

It is easy to see that u is a linear combination of rooted trees and therefore its coproduct should be a linear combination of coproducts. It means that when each component of the above sum is primitive, u is primitive. For example, coproduct of $t_1 \circ z_1$ contains non-trivial terms that induced by admissible cuts on rooted trees t_1, z_1 and also admissible cuts on $t_1 \circ z_1$. If $t_1 \circ z_1$ is primitive, then all of these terms should be canceled.

Lemma 7.1.15. *For a given primitive element $u \in \frac{\mathbb{K}[T(A)]}{I_\pi}$, t_1, t_2, z_1, z_2, s are empty tree.*

Corollary 7.1.16. $u = t + i \in \frac{\mathbb{K}[T(A)]}{I_\pi}$ *is primitive iff i be the empty tree and t primitive.*

With attention to the Hall set corresponding to H_U and the related Lyndon words, a new characterization from primitives at this level can be given.

Lemma 7.1.17. *Let $\mathbf{H}(\mathbf{T(A)})_U$ be the corresponding Hall set to the Lyndon words on the locally finite total order set $A = (f_n)_{n \in \mathbb{N}}$. $t \in \mathbf{H}(\mathbf{T(A)})_U$ is primitive iff $t = B_a^+(\mathbb{I})$ for some $a \in A$.*

It means that H_U has just primitive elements of degree zero (i.e. a vertex labeled by one arbitrary generator f_n) and its related one cocycle is denoted by $B_{f_n}^+$. In fact, lemma 7.1.17 is a representation of this note that a (Lyndon) word w is primitive if and only if $w = a$ for some $a \in A$. This story introduces generators (i.e. one cocycles) of $HH_\epsilon^1(H_U)$ corresponding to primitive elements of this Hopf algebra. On the other hand, theorem 7.1.2 makes clear the Hochschild cohomology of H_U at the Lie algebra level and it means that

$$H^n_{Lie}(L_U, U(L_U)) \cong HH^n(U(L_U), U(L_U)). \tag{7.1.9}$$

Corollary 7.1.18. *Lie polynomials are exactly the primitives in the graded dual of H_U and therefore for each Lie polynomial, one can identify its corresponding one cocycle at the level of the Lie algebra cohomology.*

7.2 Universal Hopf algebra of renormalization and factorization problem

If we look at to the structure of the universal affine group scheme $\mathbb{U}$ and its corresponding Hopf algebra $H_\mathbb{U}$, then it can be understood that this Hopf algebra is free from any physical dependency. In other words, it is independent of all renormalizable theories and this property separates this specific Hopf algebra from other renormalization Hopf algebras of Feynman diagrams of theories. In this part, with attention to the combinatorics of Dyson-Schwinger equations and on the basis of factorization problem of Feynman graphs, we are going to improve the independency of the specific Hopf algebra $H_\mathbb{U}$ to the level of nonperturbative study. This property leads us to remove the uniqueness problem of factorization and also, it plays an essential role to consider the behavior of the Connes-Marcolli categorical approach with respect to these equations where as the consequence, we will explain in the next section that the Connes-Marcolli's universal category $\mathcal{E}$ could preserve its universality at the level of Dyson-Schwinger equations.

Kreimer discovered a very trickly interaction between Dyson-Schwinger equations and Euler products and in this process he applied one important concept in physics namely, factorization. Factorization of Feynman diagrams into primitive components is in fact a brilliant highway for physicist to transfer information from perturbative theory (as a well-defined world) to nonperturbative theory (as an ill-defined world). This conceptional translation can be explained by Dyson-Schwinger equations and we know that they are equations which formally solved in an infinite series of graphs. On the other hand, it was shown that Feynman diagrams can be decomposed to primitive graphs with bidegree one in a recursive procedure such that the extension of this mechanism to the level of DSEs is known as one important problem. Fortunately, with introducing a new technical shuffle type product on 1PI Feynman graphs, one can find the answer of this question. One can rewrite the solution of a DSE based on this new shuffle type product. The surprising note is that Euler factorization and Riemann ζ−function play a large role in this process. [54, 55, 56, 57, 59, 62]

With help of the given rooted tree reformulation of the universal Hopf algebra of renormalization and based on the combinatorial approach in the study of DSEs, we are going to consider this family of equations at the level of $H_\mathbb{U}$ and after that with respect to the uniqueness problem of factorization of Feynman diagrams, we shall see that how one can extend the universal property of this Hopf algebra to nonperturbative theory. This result can provide remarkable requirements to find a new geometric interpretation from DSEs underlying a categorical configuration. [97, 99]

Definition 7.2.1. *For each labeled rooted tree t, the finite value*

$$w(t) := \sum_{v \in t^{[0]}} |dec(v)|$$

is called decoration weight of t.

Theorem 7.2.1. *For a given equation DSE in the Hopf algebra H_{CK}, there is an explicit presentation from the generators c_n, ($n \in \mathbb{N}$) (identified with the unique solution of DSE) at this level. We have*

$$c_0 = \mathbb{I}, \quad c_n = \sum_{t, w(t)=n} \frac{t}{|sym(t)|} \prod_{v \in t^{[0]}} \rho_v$$

such that
$$\rho_v = \omega_{|dec(v)|} \frac{(|dec(v)|+1)!}{(|dec(v)|+1-fer(v))!},$$
for the case $fert(v) \le |dec(v)|+1$ *and* $\rho_v = 0$, *for otherwise.* [8]

Corollary 6.2.27 allows us to lift this theorem to the level of H_U.

Proposition 7.2.2. *For a given combinatorial equation DSE in* H_U, *its unique solution* $c = (c_n)_n$ *is determined by*
$$c_0 = I_\pi,$$
$$c_n = \sum_{t \in \mathbf{H}(\mathbf{T}(\mathbf{A}))_U, t \notin I_\pi, w(t) = n} t \prod_{v \in t^{[0]}} \rho_v + I_\pi.$$

Proof. It is proved by proposition 7.1.1, theorem 7.2.1 and this note that Hall trees (forests) have no symmetries. □

Let $\mathbb{K}[[h]]$ be the ring of formal series in one variable over $\mathbb{K}$. Foissy in [35, 36] considers some interesting classes of DSEs (given by elements of $\mathbb{K}[[h]]$) in Hopf algebras of rooted trees and also, he classifies systematically their associated Hopf subalgebras. One can lift his results to the level of labeled rooted trees and hence H_U.

Definition 7.2.3. *The composition of formal series gives a group structure on the set*
$$G := \{h + \sum_{n \ge 1} a_n h^{n+1} \in \mathbb{K}[[h]]\}.$$

Hopf algebra of functions on the opposite of the group G *is called Faa di Bruno Hopf algebra. It is a connected graded commutative non-cocommutative Hopf algebra and denoted by* H_{FdB} *such that for each* $f \in H_{FdB}$ *and* $P, Q \in G$, *its coproduct is given by*
$$\Delta(f)(P \otimes Q) = f(Q \circ P).$$

Remark 7.2.4. *One can show that* H_{FdB} *is the polynomial ring in variables* Y_i *(*$i \in \mathbb{N}$*)*, *where*
$$Y_i : G \longrightarrow \mathbb{K}, \quad h + \sum_{n \ge 1} a_n h^{n+1} \longmapsto a_i.$$

Proposition 7.2.5. *Let* $\mathbb{K}[[h]]_1$ *be the set of elements in* $\mathbb{K}[[h]]$ *with constant term* 1 *and* $P \in \mathbb{K}[[h]]_1$.
(i) Dyson-Schwinger equation $X_P = B^+_{f_n}(P(X_P))$ *in* $H_U[[h]]$ *has a unique solution with the associated Hopf subalgebra* $H_U^{\alpha,\beta}(P)$ *iff there exists* $(\alpha, \beta) \in \mathbb{K}^2$ *such that*
$$(1 - \alpha\beta h)P'(h) = \alpha P(h), \quad P(0) = 1$$

(ii) For $\beta \ne -1$, $H_U^{1,\beta}(P)$ *is isomorphic to the quotient Hopf algebra* $\frac{H_{FdB}}{I_\pi}$.
(iii) $H_U^{1,-1}(P)$ *is isomorphic to a quotient Hopf algebra* $\frac{SYM}{J}$ *of Hopf algebra of symmetric functions.*
(iv) $H_U^{0,1}(P)$ *is isomorphic to the quotient Hopf algebra* $\frac{\mathbb{K}[\bullet]}{I_\pi}$.

Proof. It is proved based on the rooted tree version of H_U and the given results in sections 3, 4, 5 in [35]. For the third part, there is a homomorphism $\theta : SYM \longrightarrow H_{CK}$ that sends each generator $m_{\underbrace{(1,...,1)}_{n}}$ to the ladder tree l_n. Set $J := \theta^{-1}(I_\pi)$. □

Moreover, Hoffman suggests a new procedure to study DSEs with translating equations to a quotient of noncommutative version of the Connes-Kreimer Hopf algebra namely, Foissy Hopf algebra. One can improve his main result to our interesting level.

Proposition 7.2.6. *The unique solution of the equation*

$$X = I_\pi + B^+_{f_n}(X^p)$$

in H_U where $p \in \mathbb{R}$ is determined by

$$t_n = \sum_{t \in \mathbf{H}(\mathbf{T}(\mathbf{A}))_U} e(t) C_p(t) t$$

such that $e(t)$ is the number of Hall planar rooted trees s such that $\alpha_2(s) = t$ (defined in theorem 2.2.5).

Proof. The unique solution of DSE is given by a formal sum

$$X = I_\pi + t_1 + t_2 + ...$$

such that t_n is a Hall tree in $\mathbf{H}(\mathbf{T}(\mathbf{A}))_U$ with degree n. Set

$$\widetilde{X} := t_1 + t_2 + ...$$

Equation DSE can be changed to the form

$$\widetilde{X} = B^+_{f_n}((I_\pi + \widetilde{X})^p).$$

Since the operator $B^+_{f_n}$ increases degree, it is easy to see that

$$t_{n+1} = B^+_{f_n}(\{I_\pi + \binom{p}{1}\widetilde{X} + \binom{p}{2}\widetilde{X}^2 + ...\}_n)$$

such that $\{\}_n$ is the component of degree n. There is a natural homomorphism $\alpha_2 : H_F \longrightarrow H_{CK}$ that maps each planar rooted tree to its corresponding rooted tree without notice to the order in products. One can lift this homomorphism to the level of labeled rooted trees and apply it to study the given DSE at the level of the quotient Hopf algebra $\frac{H_F(A)}{I_\pi}$. Let

$$I_\pi + \widetilde{Y} = I_\pi + s_1 + s_2 + ...$$

be its solution in this new level such that s_n is a Hall planar rooted tree of degree n in $\mathbf{H}(\mathbf{T}(\mathbf{A}))_U$. Induction shows that

$$s_n = \sum_{t \in \mathbf{H}(\mathbf{P}_{n-1}(\mathbf{A}))_U} C_p(t) t$$

such that

$$C_p(t) = \prod_{v \in \overline{V}(t)} \binom{p}{c(v)}$$

where $c(v)$ is the number of leaves of v, $\overline{V}(t)$ is the set of vertices of t with $c(v) \neq 0$ and $\mathbf{H}(\mathbf{P_{n-1}(A)})_U$ is the Hall subset of $\mathbf{H}(\mathbf{T(A)})_U$ generated by planar rooted trees of degree n. It is observed that for each Hall planar rooted tree s, $C_p(s) = C_p(\alpha_2(s))$. Since $\alpha_2(\widetilde{Y}) = \widetilde{X}$, according to the theorem 6.2 in [44], one can obtain a clear presentation from the unique answer of DSE. □

Now it is the time to consider the attractive relation between Dyson-Schwinger equations and Euler products underlying the factorization problem.

Definition 7.2.7. *The analytic continuation of the sum $\sum_n \frac{1}{n^s}$ is called Riemann ζ-function. If $\Re(s) > 1$, then it has an Euler product over all prime numbers given by $\zeta(s) = \prod_p \frac{1}{1-p^{-s}}$.*

Definition 7.2.8. *For a sequence of primitive 1PI graphs $J = (\gamma_1, ..., \gamma_k)$ in the renormalizable theory Φ, a Feynman graph Γ is called compatible with J (i.e. $\Gamma \sim J$), if*

$$< Z_{\gamma_1} \otimes ... \otimes Z_{\gamma_k}, \Delta^{k-1}(\Gamma) >= 1.$$

Lemma 7.2.9. *Let n_Γ be the number of compatible sequences with Γ. Define a product on 1PI graphs given by*

$$\Gamma_1 \uplus \Gamma_2 := \sum_{I_1 \sim \Gamma_1, I_2 \sim \Gamma_2} \sum_{\Gamma \sim I_1 \star I_2} \frac{1}{n_\Gamma} \Gamma.$$

It is a commutative associative product.

Lemma 7.2.10. *Define a relation*

$$\Gamma_1 \leq \Gamma_2 \iff < Z_{\Gamma_1}^-, \Gamma_2 > \neq 0.$$

It determines a partial order relation based on subgraphs.

Remark 7.2.11. *(i) One can rewrite the Connes-Kreimer coproduct by*

$$\Delta(\Gamma) = \sum_{\Gamma_1, \Gamma_2} \zeta(\Gamma_1, \Gamma_2) \Gamma_1 \otimes \Gamma_2$$

where if $\Gamma_1 \leq \Gamma_2$, then $\zeta(\Gamma_1, \Gamma_2) = 1$ and otherwise $\zeta(\Gamma_1, \Gamma_2) = 0$.
(ii) The product $\uplus$ is a generalization of the shuffle product $\star$ and it means that $\star$ appropriates for totally ordered sequences whenever $\uplus$ is just for partial order relation (on subgraphs). [54, 55]

With the help of defined shuffle type product on Feynman diagrams, Kreimer shows that Euler products exist in solutions of DSEs.

Lemma 7.2.12. *Unique solution of the equation*

$$X = \mathbb{I} + \sum_\gamma \alpha^{k_\gamma} B_\gamma^+(X^{k_\gamma})$$

(such that k_γ is the degree of γ) has an $\uplus-$Euler product given by

$$X = \prod_\gamma^\uplus \frac{1}{1 - \alpha^{k_\gamma}\gamma}.$$

[54, 55, 57]

Remark 7.2.13. *The uniqueness of this factorization in gauge theories is lost and for removal this problem, we should work on the quotient Hopf algebras. [55, 57, 101, 102, 103]*

Now we know that uniqueness is the main problem of factorization and in continue we want to focus on this lack to show the advantage of the universal Hopf algebra of renormalization.

Theorem 7.2.2. *Consider set H_{pr} of all sequences $(p_1, ..., p_k)$ of prime numbers such that the empty sequence is denoted by 1 and define a map B_p^+ such that its application on a sequence $J = (p_1, ..., p_k)$ is the new sequence $(p, p_1, ..., p_k)$. There is a Hopf algebra structure on H_{pr} such that its compatible coproduct with the shuffle product is determined by*

$$\Delta(B_p^+(J)) = B_p^+(J) \otimes 1 + [id \otimes B_p^+]\Delta(J),$$

$$\Delta(1) = 1 \otimes 1,$$

$$\Delta((p)) = (p) \otimes 1 + 1 \otimes (p).$$

$$B_{p_1}^+(J_1) \star B_{p_2}^+(J_2) = B_{p_1}^+(J_1 \star B_{p_2}^+(J_2)) + B_{p_2}^+(B_{p_1}^+(J_1) \star J_2).$$

[60]

Remark 7.2.14. *It is natural to think that the operator B_p^+ is a Hochschild one cocycle and it means that with this operator one can introduce Dyson-Schwinger equations at the level of H_{pr}.*

Consider Dyson-Schwinger type equation

$$X(\alpha) = 1 + \sum_p \alpha B_p^+[X(\alpha)] \qquad (7.2.1)$$

in H_{pr}. It has a decomposition given by

$$X(\alpha) = \prod_p^\star \frac{1}{1 - \alpha(p)}. \qquad (7.2.2)$$

Because of the shuffle nature of this Hopf algebra, one can suggest to apply universal Hopf algebra of renormalization. It is observed that for finding the Euler factorization, one should define a new product on Feynman diagrams of a theory and on the other hand, the uniqueness of this factorization is not available in general. Now we see that with working at the level of H_U, this indicated problem can be solved.

Proposition 7.2.15. *There exists an Euler factorization in the universal Hopf algebra of renormalization.*

Proof. There are two shuffle structures to apply in H_U for receiving factorization. One of them namely, $\star$ is exactly the same as the shuffle product on H_U. One can show that
(i) The product of H_U is integral,
(ii) There is a combinatorial $\star$—Euler product (comes from a class of DSEs) in the universal Hopf algebra of renormalization.

For the second claim, it is clear that for each variable f_n ($n \in \mathbb{N}_{>0}$) in H_U, $B_{f_n}^+$ is a Hochschild one cocycle. Consider equation

$$X(\alpha) = 1 + \sum_{f_n} \alpha B_{f_n}^+[X(\alpha)]$$

in H_{U}. By replacing prime numbers with these variables in (7.2.2) and also with notice to the shuffle structure in H_{U}, one can get the decomposition

$$X(\alpha) = \prod_{f_n}^{\star} \frac{1}{1 - \alpha(f_n)}$$

such that (f_n) is a word with length one. □

It was shown that an extension of the shuffle product can be applied to obtain a factorization for the formal series of Feynman diagrams of the solution of a DSE. For the word $w = f_{k_1}...f_{k_n}$ in H_{U}, a word v is called compatible with w (i.e. $v \sim w$), if

$$< p_{f_{k_1}} \otimes ... \otimes p_{f_{k_n}}, \Delta^{k_n - 1}(v) >= 1 \quad (7.2.3)$$

such that $p_{f_n} \equiv e_{-n}$s are Hall polynomials of primitive elements f_ns. (One should stress that Hall polynomials form a basis (at the vector space level) for the Lie algebra). Therefore

$$v \sim w \iff < e_{-k_1} \otimes ... \otimes e_{-k_n}, \Delta^{k_n - 1}(v) >= 1. \quad (7.2.4)$$

Let n_w be the number of compatible words with w. It determines our interesting product

$$w_1 \uplus w_2 := \sum_{v_1 \sim w_1, v_2 \sim w_2} \sum_{w \sim v_1 \star v_2} \frac{1}{n_w} w. \quad (7.2.5)$$

Now consider Dyson-Schwinger equation

$$X = 1 + \sum_{f_n} \alpha^n B_{f_n}^+(X^n) \quad (7.2.6)$$

in the universal Hopf algebra of renormalization. Its unique solution has an $\uplus$−Euler product given by

$$X = \prod_{f_n}^{\uplus} \frac{1}{1 - \alpha^n f_n}. \quad (7.2.7)$$

Corollary 7.2.16. *There is a unique factorization into the Euler product in the universal Hopf algebra of renormalization.*

Proof. With attention to the given coproduct in theorem 6.1.1 and also proposition 7.2.15, it can be seen that

$$v \sim w \iff v = w \implies n_w = 1 \implies w_1 \uplus w_2 = w_1 \star w_2.$$

Therefore $\uplus$−Euler product and $\star$−Euler product in the universal Hopf algebra of renormalization are the same. On the other hand, we know that each word has a unique representation with a decreasing decomposition to Hall polynomials and moreover, corollary 6.2.44 shows that these elements determine a basis at the vector space level for the free algebra H_{U}. □

At last, let us consider the possibility of defining the Riemann ζ−function in H_{U}.

Proposition 7.2.17. *The Riemann ζ−function can be reproduced from a class of DSEs in the universal Hopf algebra of renormalization.*

Proof. One can define an injective homomorphism from H_{prime} to H_{U} given by

$$J = (p_1, ..., p_n) \longmapsto w_J = f_{p_1}...f_{p_n}.$$

There is an interesting notion (in [54]) to obtain Riemann ζ-function from a class of DSEs in H_{prime} such that one can lift it to the level of H_{U}. Consider the equation

$$X(\alpha) = 1 + \sum_p \alpha B^+_{f_p}[X(\alpha)]$$

in H_{U} such that the sum is on prime numbers. Choose a homomorphism ϕ_s of H_{U} given by

$$\phi_s(w) = \frac{1}{|w|!} pr(w)^{-s}$$

such that for the word $w = f_{k_1}...f_{k_n}$, $pr(w) := k_1...k_n$. It is observed that

$$lim_{\alpha \longrightarrow 1} \phi_s[X(\alpha)] = \zeta(s).$$

The Euler product of the above DSE is given by

$$X(\alpha) = \prod_p^\star \frac{1}{1 - \alpha(f_p)}$$

and one can see that

$$\phi_s(\prod_p^\star \frac{1}{1-\alpha(f_p)}) = \prod_p \frac{1}{1-p^{-s}} = \zeta(s).$$

□

So the Riemann-Hilbert correspondence can help us to recover the factorization problem which means that this explained procedure yields one essential reason to generalize the concept of universality of H_{U} in nonperturbative theory.

Corollary 7.2.18. *The mentioned shuffle type Euler factorization can be produced by the original product of H_{U} which insures the uniqueness problem. Universal Hopf algebra of renormalization can preserve its independency from physical theories at the level of Dyson-Schwinger equations.*

7.3 DSEs in a categorical framework

We discussed that how combinatorial Dyson-Schwinger equations can determine a systematic formalism to consider nonperturbative theory based on the Connes-Kreimer perturbative renormalization. It means that this Hopf algebraic reinterpretation can lead to an extremely practical strategy to discover some unknown parts of the theory of quantum fields. In this process we understood that from each DSE one can associate a Hopf subalgebra of Feynman diagrams of a fixed theory. In this part we want to work on these Hopf subalgebras and introduce a new framework to study Dyson-Schwinger equations based on the Connes-Marcolli's universal approach. This purpose can be derived by finding an interrelationship between these equations and objects of the category of flat equi-singular vector bundles where it has an essential universal property in the mathematical treatment of the perturbative renormalization. In this generalization one can characterize a new family of equations (i.e. *universal Dyson-Schwinger equations*) such that it establishes a new procedure to calculate Feynman integrals in the sense that at first, one can find the solution of an equation at the simplified universal level (i.e. DSEs in H_U) and then with using graded representations (introduced by Connes-Marcolli theory), we will project this solution to the level of an arbitrary renormalizable theory. In addition, by this way, one can find an arterial road from combinatorial DSEs to the universal category $\mathcal{E}$ where as the consequence, it mentions a new geometric interpretation from Dyson-Schwinger equations underlying the Riemann-Hilbert correspondence. So it makes possible to expand the universality of the category $\mathcal{E}$ to the level of these equations. [97, 99]

The universality of the category $\mathcal{E}$ is determined by its relationship with categories connected with renormalizable theories and it was shown in theorem 5.2.2 that this neutral Tannakian category enables to cover categories of all renormalizable theories as full subcategories and moreover, it determines the specific affine group scheme $\mathbb{U}^*$. Here we are going to apply Hopf subalgebra structures depended on Dyson-Schwinger equations to find a relation between these equations and objects of $\mathcal{E}$.

Let us consider the Dyson-Schwinger equation DSE in $H[[\alpha]]$ with the associated Hopf subalgebra H_c and affine group scheme G_c and let $\mathfrak{g}_c$ be its corresponding Lie algebra.

Lemma 7.3.1. *Generators of the Lie algebra $\mathfrak{g}_c$ are linear maps*

$$Z_n : H_c \longrightarrow \mathbb{C}, \quad Z_n(c_l) = \delta_{n,l}.$$

Lemma 7.3.2. *One can make a trivial principal G_c–bundle $P_c = B \times G_c$ over the base space $B = \Delta \times \mathbb{G}_m(\mathbb{C})$ such that its restriction on B^0 is denoted by $P_c^0 = B^0 \times G_c$.*

Corollary 7.3.3. *(i) Equivalence relation given by definition 5.1.3 provides a bijective correspondence between minus parts of the Birkhoff decomposition of loops (with values in G_c) and elements of the Lie algebra $\mathfrak{g}_c$.*
(ii) Theorem 5.1.3 guarantees the existence of the classes of equi-singular flat connections with respect to the elements of this Lie algebra. Therefore a classification of equi-singular flat connections on the vector bundle connected with the combinatorial equation DSE in the theory Φ will be determined.

Corollary 7.3.4. *For a given equation DSE in the theory Φ, equivalence classes of flat equi-singular G_c–connections on P_c^0 are represented by elements of the Lie algebra $\mathfrak{g}_c$ and also, each element of this Lie algebra identifies one specific class of equi-singular G_c–connections. This process is done independent of the choice of a local regular section $\sigma : \Delta \longrightarrow B$ with $\sigma(0) = y_0$.*

Proposition 7.3.5. *Let $c = (c_n)_{n \in \mathbb{N}}$ be the unique solution of DSE. Result 7.3.4 shows that for each $k \in \mathbb{N}$ there exists a unique class of flat equi-singular connections ω_c^k on P_c^0 such that*

$$\omega_c^k \sim \mathbf{D}\gamma_{Z_k},$$

$$\gamma_{Z_k}(z,v) = Te^{-\frac{1}{z}\int_0^v u^Y(Z_k)\frac{du}{u}}, \quad u = tv, \ t \in [0,1].$$

Let V^l be an arbitrary l–dimensional vector space generated by some elements of $(c_n)_{n \in \mathbb{N}}$ and $\psi_c^l : G_c \longrightarrow Gl(V^l)$ be a graded representation. By theorem 5.2.3, the pair (ω_c^k, ψ_c^l) identifies an element from the category of flat equi-singular vector bundles $\mathcal{E}$.

Corollary 7.3.6. *For a given Dyson-Schwinger equation DSE in the theory Φ, a family of objects of the category $\mathcal{E}$ will be determined.*

When we consider the Riemann-Hilbert correspondence underlying the Connes-Kreimer theory of perturbative renormalization, the universality of the category $\mathcal{E}$ is characterized by this fact that $\mathcal{E}$ carries a geometric representation from all renormalizable theories as subcategories. Now this notion can be expanded to the level of combinatorial DSEs.

Definition 7.3.7. *For a fixed equation DSE in the theory Φ, one can define a subcategory $\mathcal{E}_c^\Phi$ of $\mathcal{E}$ such that its objects are introduced by corollary 7.3.6.*

Proposition 7.3.8. *For each given equation DSE, there are classes of flat equi-singular G_c-connections such that they introduce a category. Instead of working on this category, one can go to a universal framework and concentrate on a full subcategory of $\mathcal{E}$ of those flat equi-singular vector bundles that are equivalent to the finite dimensional linear representations of G_c^*. It provides this fact that $\mathcal{E}_c^\Phi$ has power to store a geometric description from DSE.*

With attention to the algebro-geometric dictionary (in the minimal subtraction scheme in dimensional regularization [19]), we know that each loop γ_μ in the space of diffeographisms can associate a homomorphism $\phi_{\gamma_\mu} : H \longrightarrow \mathbb{C}$ and then we can perform Birkhoff decomposition at the level of these homomorphisms to obtain physical information.

Lemma 7.3.9. *There is a surjective map from the affine group scheme G to the affine group scheme G_c.*

Proof. We know that H_c is a Hopf subalgebra of H. With restriction one can map each element ϕ in the complex Lie group scheme $G(\mathbb{C})$ to its corresponding element $\phi_c \in G_c(\mathbb{C})$. On the other hand, there is an injection from H_c to H such that it determines an epimorphism from $Spec(H)$ to $Spec(H_c)$. □

Proposition 7.3.10. *Objects of the category $\mathcal{E}_c^\Phi$ store some parts of physical information of the theory with respect to the Dyson-Schwinger equation DSE.*

This fact shows that with help the of objects of the category of flat equi-singular vector bundles, a geometrically analysis from all of the combinatorial DSEs in a given theory can be investigated and since according to [60] these equations address nonperturbative circumstances, therefore category $\mathcal{E}$ can preserve its universal property at this new level.

It was shown that Hopf algebras of renormalizable physical theories and H_U have the same combinatorial source (namely, rooted trees). By applying the next fact, one can find an interesting idea to compare Dyson-Schwinger equations at the level of renormalizable physical theories with their corresponding at the level of the universal Hopf algebra of renormalization.

We know that with the help of Hochschild one cocycles (identified by primitive elements of the Hopf algebra), one can characterize combinatorial DSEs. Fix an equation DSE in H with the associated Hopf subalgebra H_c. According to theorem 5.2.3, for the equivalence class of flat equi-singular connections ω on $\widetilde{P}^0$ one can identify a graded representation ρ_ω. Let ω_c be the flat equi-singular connection on $\widetilde{P}_c^0 := B^0 \times G_c^*$ corresponding to ω with the associated graded representation ρ_{ω_c}. On the other hand, one can consider DSE in a decorated version of the

Connes-Kreimer Hopf algebra of rooted trees. Theorem 7.1.1 provides an explicit reformulation from generators of the Hopf algebra H_c such that by proposition 7.2.2, we can lift these generators to the level of the rooted tree representation of $H_{\mathbb{U}}$. This process determines a new equation DSE_u and a Hopf subalgebra H_u of $H_{\mathbb{U}}$ with the related affine group scheme $\mathbb{U}_c$.

Lemma 7.3.11. *For a fixed equi-singular flat connection ω, one can provide a graded representation $\rho_{\omega_c}^c : \mathbb{U}_c^* \longrightarrow G_c^*$ such that it is a lift of the representation ρ_ω and characterized with ρ_{ω_c}. In summary, we have the following commutative diagram.*

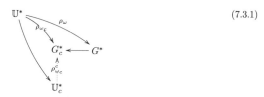

(7.3.1)

Proposition 7.3.12. *The morphism $\rho_{\omega_c}^c$ induced by representations ρ_ω, ρ_{ω_c} provides the concept of universal Dyson-Schwinger equations and it means that DSE_u maps to DSE (or DSE lifts to DSE_u) under the representation $\rho_{\omega_c}^c$.*

When $\mathbb{K} = \mathbb{Q}$, the fiber functor $\varphi : \mathcal{E}_{\mathbb{Q}} \longrightarrow \mathcal{V}_{\mathbb{Q}}$ is given by $\varphi = \bigoplus \varphi_n$ such that for each element Θ of the category,

$$\varphi_n(\Theta) := Hom(\mathbb{Q}(n), Gr_{-n}^W(\Theta)). \qquad (7.3.2)$$

For each n, $\mathbb{Q}(n) = [V, \triangledown]$ is an object in $\mathcal{E}_{\mathbb{Q}}$ such that V is an one dimensional $\mathbb{Z}$-graded $\mathbb{Q}$-vector space placed in degree n and $\triangledown = d$ (i.e. ordinary differentiation in one variable). We explained that how one can identify some elements of this category with objects given in corollary 7.3.6 and now it would be remarkable to see that elements $\mathbb{Q}(n)$ are represented with respect to a given DSE and for this work, the class of trivial connections should be calculated. Because proposition 1.101 in [22] provides this fact that the connection $\triangledown$ identifies a connection ω where $\triangledown = d + \omega$. In other words, we have to find an element in the Lie algebra $\mathfrak{g}_c$ such that its corresponding equi-singular connection is equivalent to 0 and one can show that Z_0 plays this role. We have

$$\theta_{-t}(Z_k) = e^{-tk} Z_k \qquad (7.3.3)$$

$$k = 0 \Longrightarrow \theta_{-t}(Z_0) = Z_0 : H_c \longrightarrow \mathbb{C}, \quad Z_0(c_k) = \delta_{0,k}. \qquad (7.3.4)$$

By theorem 1.60 in [22], for the element $\beta = Z_0$ in $\mathfrak{g}_c$, we have

$$\gamma_-^c(z) = Te^{-\frac{Z_0}{z} \int_0^\infty dt}. \qquad (7.3.5)$$

On the other hand, we know that the expression $Te^{\int_0^u \alpha(t)dt}$, such that $\alpha(t) = 1$ for $t = t_0$ and $\alpha(t) = 0$ for otherwise, is the value $g(u)$ of the solution for the equation

$$dg(u) = g(u)\alpha(u)du, \quad g(0) = 1. \qquad (7.3.6)$$

Therefore by (7.3.6), it can be seen that

$$g(t) = const. \Longrightarrow \mathbf{D}g = 0 \Longrightarrow \omega \sim 0. \qquad (7.3.7)$$

Corollary 7.3.13. *One can represent elements $\mathbb{Q}(n)$ with a given equation DSE.*

Proof. Let $c = (c_n)_{n \in \mathbb{N}}$ be the unique solution of a fixed equation DSE and V be the one dimensional $\mathbb{Q}$-vector space generated by c_n placed in degree n. It is observed that $\omega_0 \sim D\gamma_{Z_0}^c$ such that $\gamma_{Z_0}^c$ is the constant loop and it means that $\omega_0 = 0$. Since $\triangledown = d + \omega_0$, the proof is complete. $\square$

One should note that there are different choices to define the vector space V in the representation of $\mathbb{Q}(n)$ with a fixed DSE but all of them belong to the isomorphism class of one dimensional $\mathbb{Z}$-graded $\mathbb{Q}$-vector spaces. For a given equation DSE, result 7.3.13 shows that the equation (7.3.2) can play the role of the fiber functor for the full abelian tensor category $\mathcal{E}_c^\Phi$.

Chapter 8

From Combinatorial Dyson-Schwinger Equations to the Category of Feynman Motivic Sheaves

Problem about the existence of a universal cohomological theory for algebraic varieties defined over a base field k (and taking values into an abelian tensor category) is actually well known as the initial point for introducing motives in algebraic geometry. Theory of motives leads to develop a unified setting underlying several cohomological theories such as Betti, de Rham, l−adic, crystalline and etale. In this machinery, to construct an abelian tensor category is the main target which provides a linearization of the category of algebraic varieties such that the resulting category encodes standard conjectures of Grothendieck. The construction of a category of motives (mixed motives) related to general varieties is a difficult procedure and further, the noncommutative version of motivic objects provides the motivic cohomology applied in the construction of a universal cohomological theory. The structure of the category $\mathcal{E}$ leads us to find out mixed Tate motives in the study of Quantum Field Theory which means that Hopf algebraic treatment indicates a category of mixed Tate motives associated to each physical theory. [82, 22, 6]

In this section, we are going to apply the Aluffi-Marcolli motivic approach to quantum field theory ([3, 4, 81, 82, 83]) to obtain a new motivic investigation in the study of DSEs. Following this purpose, we will use the mentioned geometric interpretation in the previous section to produce a new family of Picard-Fuchs equations related with nonperturbative theory. In addition, we figure out the construction of a category of Feynman motivic sheaves connected with the solutions of DSEs.

8.1 DSEs in the context of the theory of motives

Let us review briefly some elementary details about hypersurface type reformulation of Feynman integrals. For a given Feynman diagram Γ in a scalar perturbative quantum field theory in spacetime dimension D with the corresponding unrenormalized Feynman integrals

$$I(\Gamma) = \int \frac{\delta(\sum_{i=1}^n \epsilon_{v,i} k_i + \sum_{j=1}^N \epsilon_{v,j} p_j)}{q_1...q_n} d^D_{k_1}...d^D k_n, \quad (8.1.1)$$

the Schwinger parameters and the technique of change of variables enable us to formulate a graph homogeneous polynomial Ψ_Γ with the property $det \ \Psi_\Gamma = b_1(\Gamma)$ which is given by the polynomial

$$\Psi_\Gamma(t) = \sum_S \prod_{e \in S} t_e \quad (8.1.2)$$

where $S \subset \Gamma^1_{int}$ of $b_1(\Gamma)$ internal edges of Γ such that the removal of all the edges belong to S leaves a connected graph. It is easy to see that this polynomial can be rewritten by the Kirchhoff polynomial such that the zero locus of the Kirchhoff polynomial

$$\widehat{X}_\Gamma = \{t \in \mathbb{A}^n : \Psi_\Gamma(t) = 0\}, \quad (8.1.3)$$

by the convention $n = |\Gamma^1_{int}|$, determines the affine graph hypersurface associated to the Feynman diagram Γ. The homogeneous property of Ψ_Γ allows us to consider the corresponding hypersurface in $\mathbb{P}^{n-1}$ which means that the graph hypersurface X_Γ related to Γ is a hypersurface in $\mathbb{P}^{n-1}$ identified by vanishing the polynomial

$$\Psi_\Gamma(t_1,...,t_n) = \sum_{T \subset \Gamma} \prod_e t_e \quad (8.1.4)$$

such that the sum is on spanning forests T and the product is on edges e which are not in T. For a fixed external momenta p, there is another homogenous polynomial $P_\Gamma(p,t) = \sum_{c \subset \Gamma} s_c \prod_{e \in c} t_e$ where the sum is over all cuts c which divide the initial graph into two connected component $\Gamma = \Gamma_1 \cup \Gamma_2$. The associated graph hypersurface to this polynomial is useful when we want to study log divergent and non-log divergent Feynman integrals. As an immediate application of this framework, one can translate Feynman integrals to the level of integrals on projective spaces. It is known that in this parametric interpretation, we consider Feynman integrals on the region of definition of the integrand which means the hypersurface complement section.

In general, for a given Feynman diagram Γ with the primitive components $\Gamma = \Gamma_1 \bigcup \Gamma_2 \bigcup ... \bigcup \Gamma_k$ there exist hypersurfaces $X_{\Gamma_1} \subset \mathbb{P}^{n_1-1}, ..., X_{\Gamma_k} \subset \mathbb{P}^{n_k-1}$ such that n_i is the number of internal edges of the sub-graph Γ_i. Applying the parametric form, the associated Feynman integral $I(\Gamma, p)$ can be rewritten with respect to a multiplicative process which means that

$$I(\Gamma, p) = I(\Gamma_1, p_1)...I(\Gamma_k, p_k). \quad (8.1.5)$$

As an application of the motivic machinery [81, 82, 83], it is shown that the original Feynman rule can be seen as a morphism to the algebra of periods that assigns

$$U(\Gamma) = [(\mathbb{A}^n \setminus \widehat{X}_\Gamma, \widehat{\sum_n}, \Psi_\Gamma^{-\frac{D}{2}} w_n, \sigma_n)] \quad (8.1.6)$$

such that $\widehat{\sum_n} = \{t \in \mathbb{A}^n : \prod t_i = 0\}$. This information provides enough knowledge to lift the Connes-Kreimer perturbative renormalization to the level of the motivic Feynman rules.

Using parametric Feynman integrals and their representations on the basis of algebraic varieties can provide a practical process for interpreting unrenormalized divergent integrals and their dimensional regularized version at the level of graph polynomials. As a rigorous advantage, it will be possible to consider parametric Feynman integrals on projective spaces where by using Radon transform in projective space, a new reformulation from Feynman integrals in terms of integrals of pullbacks of forms on a hypersurface complement in projective space will be determined. As the direct result, it is possible to calculate the dimensionally regularized parametric Feynman integrals via the Mellin transform of a Gelfand-Leray form whose Fourier transform is the oscillatory integral. This procedure supports the existence of a connection between dimensionally regularized Feynman integrals and Mellin transform of Gelfand-Leray functions which can be applied to formulate these integrals in terms of asymptotic mixed Hodge structure on the fibers of the cohomological Milnor fibration. So we can reach to a generalization of the Connes-Kreimer perturbative renormalization to the level of the mixed Hodge structures underlying the Leray regularization. Readers who would like to study more details about this construction can see [3, 4, 6, 81, 82, 83].

There is a remarkable link between regular singular Gauss-Manin connections on the cohomology of the Milnor fiber and Connes-Marcolli geometric interpretation of counterterms on the basis of flat equisingular connections on the renormalization bundle. This interrelationship together with the categorical approach to DSEs (mentioned in the previous section) allow us to modify a new family of regular singular differential equations (in the context of the Gauss-Manin connections) relevant to a given DSE.

Proposition 8.1.1. *Consider the equation DSE with the related information H_{DSE}, G_{DSE} and $\mathfrak{g}_{DSE}$. There is a family of Picard-Fuchs equations such that their solutions can determine flat equisingular $\mathfrak{g}_{DSE}(\mathbb{C})$-valued connections.*

Proof. With attention to the unique factorization problem, we can lift the equation DSE to the level of the universal Hopf algebra of renormalization and work on the corresponding equation DSE_U in H_U. Then applying the commutative diagram (7.3.1) helps us to return the results to the initial equation. So we do not worry about the unique factorization which means that one can identify a family $\{\gamma_i\}_{i \in I}$ of primitive Feynman diagrams associated to the unique solution of DSE. Consider X_{γ_i}, Y_{γ_i} as the graph hypersurfaces for γ_i with respect to the Kirchhoff polynomial Ψ_{γ_i} and the homogenous polynomial $P_{\gamma_i}(p,t)$ and let U be the motivic Feynman rules character of the physical theory. On the basis of integral transforms on projective spaces which produces the projective Radon transform $\mathcal{F}_{\sum,k}$ over the simplex $\sum$ [3, 82], the Feynman integral related to the graph γ_i can be rewritten by

$$U(\gamma_i)_\xi = \mathcal{F}_{\sum,k}(f_{\gamma_i})(\xi) = \int_{\Pi_\xi} \delta(1 - \sum_i t_i) \frac{\omega_\xi(t)}{\Psi_{\gamma_i}(t)^{\frac{D}{2}} V_{\gamma_i}(t,p)^{k - \frac{Db_1(\gamma_i)}{2}}} \tag{8.1.7}$$

such that $f_{\gamma_i}(t) = \frac{V_{\gamma_i}(t,p)^{-k + \frac{Db_1(\gamma_i)}{2}}}{\Psi_{\gamma_i}(t)^{\frac{D}{2}}}$, ξ is an $(n-k)$-frame in $\mathbb{A}^n$, Π_ξ is a generic linear subspace of dimension equal to the codimension of the singular locus of the hypersurface $X_{\gamma_i} \cup Y_{\gamma_i}$ and $V_{\gamma_i}(t,p) = \frac{P_{\gamma_i}}{\Psi_{\gamma_i}(t)}$. Its dimensional regularized version will be

$$U(\gamma_i)_\xi(z) = \int_{\sum_\xi} \frac{\omega_\xi(t)}{\Psi_{\gamma_i}(t)^{\frac{D+z}{2}} V_{\gamma_i}(t,p)^{k - \frac{(D+z)b_1(\gamma_i)}{2}}}. \tag{8.1.8}$$

This picture leads us to produce a cohomology class in $H^{dim \Pi_\xi - 1}(F_\xi)$ corresponding to the integrand of the above representation such that $F_\xi \subset \prod_\xi$ is the Milnor fiber of the hypersurface with isolated singularities $X_\xi = X \cap \prod_\xi \subset \prod_\xi$.

Applying the Mellin transform ([3, 82, 83]) provides a more practical picture from the regularized motivic Feynman integral as follows

$$F_{\gamma_i,\xi}(z) = \int \Psi_{\gamma_i}^z \alpha_\xi \qquad (8.1.9)$$

such that $\alpha_\xi = \chi_\xi P_{\gamma_i}^{b_1(\gamma_i)} \Omega_\xi$. It shows that the integral $F_{\gamma_i,\xi}$ is the Mellin transform of the Gelfand-Leray function

$$J_{\gamma_i,\xi}(z) = \int_{X_z} \frac{\alpha_\xi}{df} \qquad (8.1.10)$$

where $f = \Psi_{\gamma_i}|_{\prod_\xi}$, and $X_z = \{t : f(t) = z\}$. This class of integrals can be applied to solve the regular singular Picard-Fuchs equation

$$\frac{d}{ds}I(s) = P(s)I(s), \quad P(s)_{ij} = p_{ij}(s) \qquad (8.1.11)$$

such that the components p_{ij} are determined by the Gauss-Manin connection [83].

Choose a slicing $\prod_\xi$ enriches the Hopf algebra H_{DSE}. It can be done by the process which is the same as what happens when we equip the discrete Hopf algebra of Feynman diagrams $H_D(\mathcal{T})$ by adding the data of external momenta (Theorem 3.1.3). Letting S_{γ_i} be the manifold of planes $\prod_\xi$ in $\mathbb{A}^{|Int(\gamma_i)|}$ with the convention $dim \prod_\xi \leq codim \, Sing(X_{\gamma_i})$. It can be seen that

$$S_{\gamma_i} = \bigcup S_{\gamma_i,m} \qquad (8.1.12)$$

such that $S_{\gamma_i,m}$ is the m-dimensional planes in $\mathbb{A}^{|E_{int}(\gamma_i)|}$. Consider the space $C^\infty(S_{\gamma_i})$ of test functions on S_{γ_i} with the dual space of distributions $C_c^{-\infty}(S_{\gamma_i})$. Now enrich the Hopf algebra H_{DSE} with respect to the slicing $\prod_\xi$ to produce the new commutative Hopf algebra $\widetilde{H}_{DSE} = Sym(C_c^{-\infty}(S))$ where $S = \bigcup S_{\gamma_i}$.

Applying the Mellin transform together with the equation (8.1.10) lead us to considering the integral

$$\mathcal{F}_{\gamma_i,\xi}(z) = \int_0^\infty s^z J_{\gamma_i,\xi}(s) ds \qquad (8.1.13)$$

which admits an analytic continuation to meromorphic functions. Study $\mathcal{F}_{\gamma_i,\xi}(z)$ in a small neighborhood of $z = -D$ which has an expansion as a Laurent series with finite pole part. Using a change of variable on the complex coordinate z needs to provide a small punctured disk Δ^* around 0 which includes z. Now for a given distribution σ, define

$$\phi(\gamma_i, \sigma)(z) := \mu^{-zb_1(\gamma_i)} \sigma(\mathcal{F}_{\gamma_i,\xi}(-\frac{D+z}{2})) \qquad (8.1.14)$$

such that μ is the mass parameter. ϕ is an algebra homomorphism from $\widetilde{H}_{DSE}$ to the field of meromorphic functions at $z = 0$ which can be interpreted by a family of loops γ_μ in $Loop(\widetilde{G}_{DSE}, \mu)$. It is the place to study this family of loops in the context of the Connes-Kreimer theory to generate counterterms and renormalized values based on components of the Birkhoff factorization of γ_μ. Set

$$a_\mu(z) := (\phi \circ S) * \frac{d}{dz}\phi, \quad b_\mu(z) := (\phi \circ S) * Y(\phi) \qquad (8.1.15)$$

such that $Y = \frac{d}{dt}\theta_t|_{t=0}$ where $\{\theta_t\}_t$ is a one-parameter family of automorphisms of $\widetilde{H}_{DSE}$ generated by the grading structure. a_μ and b_μ are elements in $\Omega^1(\mathfrak{g}_{DSE}(\mathbb{C}))$ which identify our interesting equisingular connection $\omega(z, u)$.

So the solution of the regular singular Picard-Fuchs equation

$$J_{\Gamma,\xi}^{(l)}(s) + p_1(s)J_{\Gamma,\xi}^{(l-1)}(s) + ... + p_l(s)J_{\Gamma,\xi}(s) = 0, \ l = b_1(\gamma_i) \tag{8.1.16}$$

determines a flat $\mathfrak{g}_{DSE}(\mathbb{C})$-valued connection

$$\omega(z, u) = u^Y(a(z))dz + u^Y(b(z))\frac{du}{u}$$

where $a(z), b(z) \in \mathfrak{g}_{DSE}(\mathbb{C})$. The chosen k-form $\alpha_\xi = \chi_\xi P_\Gamma^l \Omega_\xi$ supports the equisingularity condition. □

From the proposition 7.3.4, we know that each equivalence class ω_{DSE} of equisingular connections determines uniquely an element in the Lie algebra $\mathfrak{g}_{DSE}(\mathbb{C})$ which means that derivations in this Lie algebra provide solutions of a family of Picard-Fuchs equations. So the graph hypersurface interpretation enables us to produce flat equisingular connections (related to DSEs) in the context of singular Picard-Fuchs equations.

Remark 8.1.2. *For the given primitive components $\{\gamma_i\}_i$, the forms α_ξ can determine a subspace of the cohomology of the Milnor fiber generated by the classes $[\frac{\alpha_\xi}{df}]$. This subspace inherits a Hodge and a weight filtration with respect to the equation DSE.*

In another direction, this graph hypersurface interpretation can give us a surprising interrelationship between DSEs and motivic sheaves which shows some motivic information hidden inside these equations. It allows us to identify logarithmic Feynman motives (as some objects of the category of mixed motives) with respect to DSEs. Consider the family $\{\gamma_i\}_i$ as primitive components of the unique solution of DSE. For each γ_i, its associated Kirchhoff polynomial Ψ_{γ_i} can be seen as the morphism from $\mathbb{A}^{|E(\gamma_i)|}\setminus\widehat{X}_{\gamma_i}$ to the multilicative affine group scheme $\mathbb{G}_m$. Applying the logarithmic motive Log in the category of mixed motives over $\mathbb{G}_m$ determines a new motive $Log_{\gamma_i} := \Psi_{\gamma_i}^*(Log)$ which is an object in the category of mixed motives over $\mathbb{A}^{|E(\gamma_i)|}\setminus\widehat{X}_{\gamma_i}$.

Proposition 8.1.3. *There is a procedure based on Kirchhoff polynomial and factorization problem to construct a category $\mathcal{E}_{DSE}^{FM}$ of Feynman motivic sheaves connected to each equation DSE which is unique up to isomorphism.*

Proof. Start with the equation DSE with the associated Hopf subalgebra H_{DSE}. We discussed in the previous parts that we can handle the unique shuffle type Euler factorization for the solution of DSE. So letting $\{\gamma_i\}_{\gamma_i}$ be the family of primitive Feynman diagrams which indicated by the unique solution. The category of Feynman motivic sheaves for the equation DSE can be constructed as a subcategory of the Arapura category of motivic sheaves over $\mathbb{G}_m$ which is generated by objects of the form

$$(\Psi_{\gamma_i}, \bigwedge \setminus (\bigwedge \cap \widehat{X}_{\gamma_i}), |E(\gamma_i)| - 1, |E(\gamma_i)| - 1) \tag{8.1.17}$$

where γ_i ranges over the Feynman graphs in the family $\{\gamma_i\}_i$ and

$$\bigwedge = \{t \in \mathbb{A}^{|E(\gamma_i)|} : \prod_j t_j = 0\} \tag{8.1.18}$$

is the union of the coordinate hyperplanes [83]. Since the factorization into primitive components is unique, therefore this category can be uniquely identified. □

Perusing this procedure reaches to a cohomological interpretation

$$H^{n-1}_{\mathbb{G}_m}(\mathbb{A}^n \backslash \widehat{X}_{\gamma_i}, \bigwedge \backslash (\bigwedge \cap \widehat{X}_{\gamma_i}), \mathbb{Q}(n-1)) \qquad (8.1.19)$$

such that $n = |E_{int}(\gamma_i)|$ [83]. As the consequence, formal series of Feynman diagrams which provide solutions of DSEs can be encoded by the cohomology theory with respect to algebraic simplex.

Chapter 9

Conclusion

In this section, we provide an overview of this monograph to address some interesting notions (related with the results of this work) in the forward of the Hopf algebraic studies of quantum field theories. Roughly speaking, this research attempts to establish new progresses in the study of Quantum Field Theory underlying the renormalization Hopf algebra and in particular, we focused on two essential problems namely, integrable systems and Dyson-Schwinger equations.

The first aim was focused on quantum integrable systems under noncommutative differential forms and perturbative renormalization. The key point in our chosen approach can be summarized in the deformation of algebras based on Nijenhuis operators where these maps are induced from regularization or renormalization schemes. Indeed, multiplicativity of renormalization can report a hidden algebraic nature inside of the BPHZ method namely, a Rota-Baxter structure such that on the basis of this property, we could introduce a new family of quantum Hamiltonian systems depending on renormalization or regularization schemes. Secondly, we studied motion integrals related to Feynman rules characters and then it was observed that how Connes-Kreimer renormalization group makes possible to obtain an infinite dimensional integrable quantum Hamiltonian system. In another direction, we considered the Baditoiu-Rosenberg approach to study a new class of integrable systems related with the Lie group of diffeographisms. In addition, based on the Bogoliubov character and the BCH formula, we found a new class of fixed point equations related to motion integrals where it yields to search more geometrical meanings inside of physical parameters.

The second aim was concentrated on the study of nonperturbative Quantum Field Theory with respect to the related combinatorial Dyson-Schwinger equations, Connes-Marcolli approach to renormalizable quantum field theories and the theory of motives. At the first step, we considered the universal Hopf algebra of renormalization H_U to produce a new combinatorial framework (on the basis of Hall sets) for the study of physical information and also, Dyson-Schwinger equations at the level of H_U. At the second step, we apply this new picture of H_U to relate Dyson-Schwinger equations with objects of the Connes-Marcolli universal category of equi-singular flat vector bundles where as the result, we could produce new geometric information hidden inside of these recursive equations. At the third step, we applied the shuffle nature of H_U to produce the unique Euler factorization for Dyson-Schwinger equations on H_U. This result leads us to extend the universality of H_U to the level of Dyson-Schwinger equations. In the fourth step, applying the Riemann-Hilbert correspondence in the study of Quantum Field Theory results two family of differential systems connected with DSEs. The first class which contains differential equations depending upon equisingular flat connections such that they are classified by elements of the Lie algebra $\mathfrak{g}_{DSE}(\mathbb{C})$. The second class which contains Picard-Fuchs equations indicated on the basis of the Kirchhoff polynomials and the equations in the first class. In addition, it was discussed that from the first class we can formulate the neutral Tannakian category $\mathcal{E}_{DSE}^{T}$ which provides a category of mixed Tate motives related to each DSE. So there should be a deep interrelationship between the Picard-Fuchs type equations generated by the proposition 8.1.1 and mixed Tate motives. In another direction, using the parametric representation of Feynman integrals and Factorization problem allowed us to identify a new Arapura type category $\mathcal{E}_{DSE}^{FM}$ with respect to the unique solution of each DSE.

And finally let us introduce some notions in the forward of the this research work which can be interesting for readers of this text.

Integrable Systems. Renormalization group determined the compatibility of the introduced integrable Hamlitonian systems with the Connes-Kreimer perturbative renormalization. It should be interesting to find another integrable Hamiltonian systems. For instance with deforming the algebra $L(H, A)$ underlying Nijenhuis maps (determined with regularization schemes), one can have this chance. On the other hand, finding a physical theory such that components of Birkhoff factorization of its associated Feynman rules character play the role of motion integrals for the character, can be very important question. We want to know that is there any specific algebro-geometric property in this kind of theories?

Integrals of motion and combinatorial DSEs. Working on this question can help us to develop our approach to integrable systems to the level of nonperturbative theory and further, it leads to discover a new description from quantum motions based on Nijenhuis type Poisson brackets. It seems that with this notion one can introduce a new interrelationship between theory of Rota-Baxter algebras and Dyson-Schwinger equations.

The application of universal DSEs. Here we introduced universal Dyson-Schwinger equations and then we just showed that with using the universal nature of the universal singular frame, one can lift DSEs in different physical theories to the level of DSE_u (i.e. as the process of comparison). There are some essential questions about the importance of this class of equations. For instance, Can we apply these equations in computations as kind of a simplified toy model? What type of physical information can be analyzed explicitly with these equations?

Motivic DSEs. Following the recent observations, the renormalization theory is modified in a motivic framework ([82, 83]) where the Aluffi-Marcolli approach to motivic Feynman rules determines a category of Feynman motivic sheaves. Finding a systematic procedure to develop this motivic machinery at the level of DSEs can provide a reasonable interpretation of these equations in the context of Picard-Fuchs equations and Arapura type category. It is indeed the start point to construct a new interpretation of nonperturbative QFT in the context of the theory of motives.

Bibliography

[1] E. Abe, *Hopf algebras*, Cambridge University Press, 1980.

[2] F.V. Atkinson, *Some aspects of Baxter's functional equation*, J. Math. Anal. Appl., Vol. 7, No. 1, 1963.

[3] P. Aluffi, M. Marcolli, *Algebro-Geometric Feynman rules*, Int. J. Geom. Meth. Mod. Phys., Vol. 8, No. 1, 203-237, 2011.

[4] P. Aluffi, M. Marcolli, *Feynman motives and deletion-contraction relations*, Contemporary Mathematics, 538 (2011), 21–64.

[5] V.I. Arnold, V.V. Kozlov, A.I. Neishtadt, *Mathematical aspects of classical and celestial Mechanics*, Springer, 1997.

[6] S. Bloch, E. Esnault, D. Kreimer, *On motives associated to graph polynomials*, Commun. Math. Phys. 267 (2006), 181–225.

[7] C. Brouder, A. Frabetti, *QED Hopf algebras on planar binary trees*, J. Alg., No. 267, 298-322, 2003.

[8] C. Bergbauer, D. Kreimer, *Hopf algebras in renormalization theory: locality and Dyson-Schinger equations from Hochschild cohomology*, Eur. Math. Soc., Zurich, 2006.

[9] C. Bergbauer, D. Kreimer, *New algebraic aspects of perturbative and non-perturbative quantum field theory*, Contribution to the proceedings of the International Congress on Mathematical Physics Rio de Janeiro, 2006.

[10] G. Baditoiu, S. Rosenberg, *Feynman diagrams and Lax pair equations*, arXiv:math-ph/0611014v1, 2006.

[11] A. Cannas da Silva, *Lectures on symplectic geometry*, Springer, 2001.

[12] J.F. Carinena, J. Grabowski, G. Marmo, *Quantum bi-Hamiltonian systems*, Int. J. Mod. Phys. A, Vol. 15, No. 30, 4797-4810, 2000.

[13] A. Connes, D. Kreimer, *Hopf algebras, renormalization and noncommutative geometry*, IHES/M/98/60, 1998.

[14] A. Connes, D. Kreimer, *Renormalization in quantum field theory and the Riemann-Hilbert problem I: the Hopf algebra structure of graphs and the main theorem*, Commun. Math. Phys., Vol. 210, No. 1, 249-273, 2000.

[15] A. Connes, D. Kreimer, *Renormalization in quantum field theory and the Riemann-Hilbert problem II: the β-function, diffeomorphisms and the renormalization group*, Commun. Math. Phys., Vol. 216, No. 1, 215-241, 2001.

[16] A. Connes, D. Kreimer, *Insertion and elimination: the doubly infinite Lie algebra of Feynman graphs*, Ann. Hen. Poin., Vol. 3, No. 3, 411-433, 2002.

[17] F. Chapoton, M. Livernet, *Relating two Hopf algebras built from an operad*, Int. Math. Res. Not. IMRN, No. 24, Art. ID rnm131, 27 pp, 2007.

[18] A. Connes, M. Marcolli, *Renormalization and motivic Galois theory*, arXiv:math.NT/0409306 v1, 2004.

[19] A. Connes, M. Marcolli, *From physics to number theory via noncommutative geormetry*, Frontiers in Number Theory, Physics, and Geometry. I, 269-347, Springer, Berlin, 2006.

[20] A. Connes, M. Marcolli, *Quantum fields and motives*, J. Geom. Phys. Vol. 56, No. 1, 55-85, 2006.

[21] A. Connes, M. Marcolli, *Renormalization, the Riemann-Hilbert correspondence and motivic Galois theory*, Frontiers in Number Theory, Physics and Geometry. II, 617-713, Springer, Berlin, 2007.

[22] A. Connes, M. Marcolli, *Noncommutative geometry, quantum fields and motives*, Colloquium Publications, Vol. 55, Amer. Math. Soc., 2008.

[23] V.G. Drinfel'd, *Hamiltonian structures on Lie groups, Lie bialgebras and the geometric meaning of the classical Yang-Baxter equations*, Soviet Math. Doklady, Vol. 27, 68-71, 1983.

[24] I. Dorfman, *Dirac structures and integrability of nonlinear evolution equations*, Nonlinear Science, Theory and Applications, John Wiley, 1993.

[25] K. Ebrahimi-Fard, *On the associative Nijenhuis relation*, Elec. J. Comb., Vol. 11, No. 1, Research Paper 38, 13 pp, 2004.

[26] K. Ebrahimi-Fard, L. Guo, *Mixable shuffles, quasi-shuffles and Hopf algbras*, J. Alg. Comb., Vol. 24, No. 1, 83-101, 2006.

[27] K. Ebrahimi-Fard, L. Guo, *Rota-Baxter algebras in renormalization of perturbative quantum field theory. universality and renormalization*, 47-105, Fields Inst. Commun., 50, Amer. Math. Soc., Providence, RI, 2007.

[28] K. Ebrahimi-Fard, L. Guo, D. Kreimer, *Integrable renormalization I: the ladder case*, J. Math. Phys., Vol. 45, No. 10, 3758-3769, 2004.

[29] K. Ebrahimi-Fard, L. Guo, D. Kreimer, *Integrable renormalization II: the general case*, Ann. Hen. Poin., Vol. 6, No. 2, 369-395, 2005.

[30] K. Ebrahimi-Fard, L. Guo, D. Manchon, *Birkhoff type decompositions and Baker-Campbell-Hausdorff recursion*, arXiv:math-ph/0602004v2, 2006.

[31] K. Ebrahimi-Fard, J.M. Gracia-Bondia, Li Guo, J.C. Varilly, *Combinatorics of renormalization as matrix calculus*, arXiv:hep-th/0508154v2, 2005.

[32] K. Ebrahimi-Fard, D. Kreimer, *The Hopf algebra approach to Feynman diagram calculations*, J. Phys. A, Vol. 38, R385-R406, 2005.

[33] K. Ebrahimi-Fard, D. Manchon, *On matrix differential equations in the Hopf algebra of renormalization*, arXiv:math-ph/0606039v2, 2006.

[34] L. Foissy, *Les algebras de Hopf des arbres enracines decores*, Bull. Sci. Math., Vol. 126, 249-288, 2002.

[35] L. Foissy, *Faa di Bruno subalgebras of the Hopf algebra of planar trees from combinatorial Dyson-Schwinger equations*, Adv. Math., Vol. 218, No. 1, 136-162, 2008.

[36] L. Foissy, *Systems of Dyson-Schwinger equations*, arXiv:0909.0358 v1, 2009.

[37] A. Frabetti, *Renormalization Hopf algebras and combinatorial groups*, arXiv: 0805.4385, 2008.

[38] V. Ginzburg V, *Lectures on noncommutative geometry*, University of Chicago, arXiv:math/0506603v1, 2005.

[39] Z. Giunashvili, *Noncommutative symplectic geometry of the endomorphism algebra of a vector bundle*, arXiv:math/0212228v1, 2002.

[40] L. Guo, *Algebraic Birkhoff decomposition and its application*, International school and conference of noncommutative geometry China, arXiv:0807.2266v1, 2007.

[41] I.M. Gelfand, D. Krob, A. Lascoux, B. Leclerc, V.S. Retakh, J.Y.Thibon, *Noncommutative symmetric functions*, Adv. Math. Vol. 112, 218-348, 1995.

[42] M. E. Hoffman, *Quasi-shuffle products*, J. Alg. Comb., Vol. 11, 49-68, 2000.

[43] M. E. Hoffman, *Combinatorics of rooted trees and Hopf algebras*, Trans. Amer. Math. Soc., Vol. 355, No. 9, 3795-3811, 2003.

[44] M. E. Hoffman, *(Non)commutative Hopf algebras of trees and (quasi)symmetric functions*, arXiv:0710.3739v1, 2007.

[45] M. E. Hoffman, *Rooted trees and symmetric functions: Zhao's homomorphism and the commutative hexagon*, arXiv:0812.2419v1 [math.QA], 2008.

[46] M. E. Hoffman, *Hopf algebras of rooted trees and the Hopf algebras of symmetric functions*, International conference on vertex operator algebras and related areas, Illinois State University, 2008.

[47] R. Holtkamp, *Comparison of Hopf algebras on trees*, Arch. Math. (Basel), Vol. 80, 368-383, 2003.

[48] R. Haag, *Local quantum physics: fields, particles, algebras*, Berlin, Germany, Springer, 1992.

[49] C. Kassel, *Quantum groups*, Graduate texts in mathematics, 155. Springer-Verlag, New York, 1995.

[50] D. Kastler, *Connes-Moscovici-Kreimer Hopf algebras*, arXiv:math-ph/0104017v1, 2001.

[51] M. Khalkhali, *From cyclic cohomology to Hopf cyclic cohomology in five lectures*, Vanderbilt University, USA, 2007.

[52] D. Kreimer, *On the Hopf algebra structure of perturbative quantum field theories*, Adv. Theor. Math. Phys., Vol. 2, No. 2, 303-334, 1998.

[53] D. Kreimer, *On overlapping divergences*, Commun. Math. Phys., Vol. 204, 669, 1999.

[54] D. Kreimer, *Unique factorization in perturbative QFT*, Nucl. Phys. Proc. Suppl., Vol. 116, 392-396, 2003.

[55] D. Kreimer, *New mathematical structures in renormalizable quantum field theories*, Ann. Phys., Vol. 303, No. 1, 179-202, 2003.

[56] D. Kreimer, *What is the trouble with Dyson-Schwinger equations?*, arXiv:hep-th/0407016v1, 2004.

[57] D. Kreimer, *Structures in Feynman graphs-Hopf algebras and symmetries*, Proc. Symp. Pure Math., Vol. 73, 43-78, 2005.

[58] D. Kreimer, *The Hopf algebra structure of renormalizable quantum field theory*, IHES preprint, 2005.

[59] D. Kreimer, *Etude for linear Dyson-Schwinger equations*, IHES/P/06/23, 2006.

[60] D. Kreimer, *Anatomy of a gauge theory*, Ann. Phys., Vol. 321, No. 12, 2757-2781, 2006.

[61] D. Kreimer, *An etude in non-linear Dyson-Schwinger equations*, arXiv:hep-th/0605096v2, 2006.

[62] D. Kreimer, *Factorization in quantum field theory: an exercise in hopf algebras and local singularities*, Frontiers in Number Theory, Physics, and Geometry. II, 715-736, Springer, Berlin, 2007.

[63] D. Kreimer, *Dyson-Schwinger equations: from Hopf algebras to number theory*, Universality and renormalization, 225-248, Fields Inst. Commun., Vol. 50, Amer. Math. Soc., Providence, RI, 2007.

[64] D. Kreimer, *Algebraic structures in local QFT*, Proceedings of the Les Houches school on structures in local quantum field theory, arXiv:1007.0341 [hep-th], 2010.

[65] D. Kreimer, R. Delbourgo, *Using the Hopf algebra structure of QFT in calculations*, arXiv: hep-th/9903249V3, 1999.

[66] Y. Kosmann-Schwarzbach, B. Grammaticos, K. M. Tamizhmani, *Integrability of nonlinear systems*, Lecture Notes in Physics, Vol. 495, 1997.

[67] Y. Kosmann-Schwarzbach, F. Magri, *Poisson-Nijenhuis structures*, Ann. Inst. Henri Poincare, Vol. 53, No. 1, 35-81, 1990.

[68] P.P. Kulish, E.K. Sklyanin, *Solutions of the Yang-Baxter equations*, J. Soviet Math., Vol. 19, 1596-1620, 1982.

[69] D. Kreimer, K. Yeats, *An etude in non-linear Dyson-Schwinger equations*, Nuclear Phys. B Proc. Suppl., Vol. 160, 116-121, 2006.

[70] D. Kreimer, K. Yeats, *Recursion and growth estimates in renormalizable quantum field theory*, Cmmun. Math. Phys., Vol. 279, No. 2, 401-427, 2008.

[71] T. Krajewski, R. Wulkenhaar, *On Kreimer's Hopf algebra structure on Feynman graphs*, arXiv:hep-th/98050998v4, 1998.

[72] G. Landi, An introduction to noncommutative spaces and their geometry, Lecture Notes in Physics. New Series m: Monographs, 51. (1997) Springer-Verlag, Berlin, ISBN: 3-540-63509-2.

[73] A. Lahiri, P. B. Pal, *A first book of quantum field theory*, Alpha Science international Ltd., 2005.

[74] J. L. Loday, M.O. Ronco, *Hopf algebra of the planar binary trees*, Adv. Math., Vol. 139, 293-309, 1998.

[75] J.L. Loday, M.O. Ronco, *Combinatorial Hopf algebras*, Clays Mathematics Proceedings, Vol. 12, arXiv: math:08100435, 2010.

[76] A. Murua, *The shuffle Hopf algebra and the commutative Hopf algebra of labelled rooted trees*, Research Report, 2005.

[77] A. Murua, *The Hopf algebra of rooted trees, free Lie algebras, and Lie series*, Found. Comput. Math., 387-426, 2006.

[78] D. Manchon, *Hopf algebras, from basics to applications to renormalization*, arXiv:math.QA/0408405, 2006.

[79] S. Majid, *A quantum groups primer*, London Mathematical Society Lecture Note Series. 292, 2002.

[80] M. Marcolli, *Renormalization for dummies*, Mathematische Arbeitstagung, MPIM, 2005.

[81] M. Marcolli, *Feynman motives*, A monograph based on courses in Caltec, Springer, 2010.

[82] M. Marcolli, *Feynman integrals and motives*, In European Congress of Mathematics, Amesterdam July 2008, European Mathematical Society, 293-332, 2010.

[83] M. Marcolli, *Motivic renormalization and singularities*, Quanta of maths: 409-458. Clay Math. Proc., 11, Amer. Math. Soc., Providence, RI (2010).

[84] I. Mencattini, D. Kreimer, *Insertion and elimination Lie algebra: the ladder case*, Lett. Math. Phys., Vol. 67, No. 1, 61-74, 2004.

[85] I. Mencattini, D. Kreimer, *The structure of the ladder insertion-elimination Lie algebra*, Commun. Math. Phys., Vol. 259, No. 2, 413-432, 2005.

[86] M. W. Milnor, J. C. Moore, *On the structure of Hopf algebras*, Ann. Math., Vol. 81, No. 2, 211-264, 1965.

[87] F. Panaite, *Relating the Connes-Lreimer and Grossman-Larson Hopf algebras built on rooted trees*, Letters in Mathematical Physics 51: 211-219, 2000.

[88] C. Reutenauer, *Free Lie algebras*, Oxford, 1993.

[89] G.C. Rota, *On the foundations of combinatorial theory. I. theory of mobius functions*, 340-368, 1964.

[90] M. Rosenbaum M, J. David Vergara, H. Quevedo, *Hopf algebra primitives in perturbation quantum field theory*, J. Geom. Phys., Vol. 49, 206-223, 2004.

[91] S.S. Schweber, *QED and the men who made it: Dyson, Feynman, Schwinger, and Tomonaga*, Princeton Series in Physics, 1976.

[92] W.R. Schmitt, *Incidence Hopf algebras*, J. Pure Appl. Alg., Vol. 96, 299-330, 1994.

[93] M. Sakakibara, *On the differential equations of the characters for the renormalization group*, Mod. Phys. Lett. A, Vol. 19, N.19, 1453-1456, 2004.

[94] M.A. Semenov-Tian-Shansky, *What is a classical r-matrix?*, Funct. Anal. Appl., Vol. 17, 259-272, 1984.

[95] M.A. Semenov-Tian-Shansky, *Integrable Systems and factorization problems. factorization and integrable systems (Faro, 2000)*, 155-218, Oper. Theory Adv. Appl., Vol. 141, Birkhuser, Basel, 2003.

[96] M. Sweedler, *Hopf algebras*, W.A.Benjamin, Inc., New York, 1969.

[97] A. Shojaei-Fard, *Categorical approach to Dyson-Schwinger equations*, Proceeding of the 40th Annual Iranian Mathematics Conference (AIMC), Sharif University of Technology, Tehran, August 2009.

[98] A. Shojaei-Fard, *Fixed point equations related to motion integrals in renormalization Hopf algebra*, Int. J. Comput. Math. Sci. (WASET), Vol. 3, No. 8, 397-402, 2009.

[99] A. Shojaei-Fard, *From Dyson-Schwinger equations to the Riemann-Hilbert correspondence*, Int. J. Geom. Meth. Mod. Phys., Vol.7, No. 4, 519-538, 2010.

[100] B. Vallette, *Homology of generalized partition posets*, J. Pure Appl. Alg. Vol. 208, 699-725, 2007.

[101] W.D. van Suijlekom, *The Hopf algebra of Feynman graphs in quantum electrodynamics*, Lett. Math. Phys., Vol. 77, No. 3, 265-281, 2006.

[102] W.D. van Suijlekom, *Renormalization of gauge fields: a Hopf algebra approach*, Commun. Math. Phys., Vol. 276, No. 3, 773-798, 2007.

[103] W.D. van Suijlekom, *Hopf algebra of Feynman graphs for gauge theories*, Conference quantum fields, periods and polylogarithms II, IHES, June 2010.

[104] M. Dubois-Violette, *Some aspects of noncommutative differential geometry*, ESI-preprint, L.P.T.H.E.-ORSAY 95/78, arXiv:q-alg/9511027v1, 1995.

[105] M. Dubois-Violette, *Lectures on graded differential algebras and noncommutative geometry*, Proceedings of the workshop on noncommutative differential geometry and its application to physics, Shonan-Kokusaimura, 1999. arXiv:math/9912017v3, 2000.

[106] G. van Baalen, D. Kreimer, D. Uminsky, K. Yeats, *The QED beta function from global solutions to Dyson-Schwinger equations*, Annals Phys. 324 (2009), 205–219 .

[107] G. van Baalen, D. Kreimer, D. Uminsky, K. Yeats, *The QCD beta function from global solutions to Dyson-Schwinger equations*, Annals Phys. 325 (2010), 300–324.

[108] I.V. Volovich, D.V. Prokhorenko, *Renormalizations in quantum electrodynamics and Hopf algebras*(Russian), translation in Proc. Steklov Inst. Math., Vol. 245, No. 2, 273-280, 2004.

[109] S. Weinzierl, *Hopf algebra structures in particle physics*, arXiv: hep-th/ 0310124v1, 2003.

[110] S. Weinzierl, *The art of computing loop integrals*, arXiv:hep-ph/0604068v1, 2006.

[111] A. S. Wightman, R. Streeter, *PCT, spin and statistics and all that*, Redwood City, USA, 1989.

[112] K. Yeats, *Growth estimates for Dyson-Schwinger equations*, Ph.D. thesis, arXiv:0810.2249v1, 2008.

◎ 编辑手记

著名数学家 E. W. Montron 曾指出：

有可能在相对短的时间里向优秀的数学专业学生有效地讲授现代物理学的任何分支,因为这些学生在数学方面是成熟的.

本书是一部英文版的涉及理论物理的数学专著,中文书名可译为《黎曼-希尔伯特问题与量子场论:可积重正化、戴森-施温格方程》.

本书的作者为阿里·索贾-法尔德(Ali Shojaei-Fard),伊朗人,在沙希德·贝赫什提大学获得博士学位.他曾在豪斯道夫数学学院,马克思·普朗克数学学院和欧文·薛定谔国际数学实验物理学院做访问学者,并在攻读博士学位期间完成了研究工作.

伊朗与中国相比虽是小国,但数学并不弱,甚至还有菲尔兹奖获得者,那就是米尔扎哈尼,首位女菲尔兹奖得主,可惜她英年早逝。

正如作者在前言中所介绍:

本著作重点研究了 Hopf 代数扰动重正化在量子场论研究中的一些最新应用. 在第一部分中,我们介绍了可积重正化形式主义,作为研究基于 Feynman 图的 Hopf 代数的可积系统的一种方法. 此外,我们考虑了一种可以替代量子可积系统的方法,该方法与重正化 Hopf 代数的无穷维复李群紧密相关. 在第二部分中,我们考虑了将 Connes-Marcolli 的通用方法扩展到非扰动量子场论研究的过程中.

第一,我们在 Hall 根树的语境中研究了重正化 H_U 的通用 Hopf 代数的新组合表示,该树可被提升到(通用)对立项的水平. 我们在组合框架中研究了这种洗牌型的通用 Hopf 代数. 第二,我们将分类构造与每个 Dyson-Schwinger 方程 DSE 相关联,DSE 可以针对连接的特殊族对此类方程进行编码. 第三,在新的 H_U 图的基础上,我们在 Dyson-Schwinger 方程的水平上证明了等奇异平面矢量束范畴的普遍性. 第四,应用上述提到的分类研究,我们确定了与每个方程 DSE 相关的 Picard-Fuchs 型方程的新族(在同构米尔诺纤维化的背景下). 此外,我们将 Feynman 运动主题束的类别(作为 Arapura 类的一个子类别)与每个方程 DSE 相关联.

关于本书的具体内容,从目录中我们可以了解个大概:

1. 导论
2. 组合对象的 Hopf 代数结构
   2.1 基本性质
   2.2 组合 Hopf 代数
3. 扰动重正化
   3.1 插入算子:从 Feymann 图的一个准李代数结构到一个 Hopf 代数
   3.2 Hopf 代数的形式主义
4. 可积重正化
   4.1 可积系统:从有限维(几何方法)到无限维
   4.2 Rota-Baxter 型代数
   4.3 量子可积系统理论
   4.4 基于重正化群的可积系统
   4.5 Baditoiu-Rosenberg 框架:标准流程的延续
   4.6 不动点定理方程
5. Connes-Marcolli 理论
   5.1 对立项的几何本质:平坦准奇异连接的类别
   5.2 通用 Tannakian 类别的构造
6. 重正化的通用 Hopf 代数
   6.1 洗牌型表示
   6.2 根树型表示
   6.3 通用对立项
7. 组合 Dyson-Schwinger 方程与 Connes-Marcolli 通用

方法

    7.1 针对重正化 Hopf 代数的量子运动

    7.2 重正化的通用 Hopf 代数与分解问题

    7.3 范畴框架中的 DSEs

8. 从组合 Dyson-Schwinger 方程到 Feynman 运动主题束的范畴

    8.1 动机理论背景下的 DSEs

9. 结论

我们先来介绍一下组合的 Hopf 代数.

- 组合 Hopf 代数 ($\mathcal{H},\zeta$) 是指域 K 上的一个分次连通的 Hopf 代数. $\mathcal{H}$ 及一个保持乘法的线性函数 $\zeta:\mathcal{H}\to K$, 称为特征. M. Aguiar, N. Bergeron 和 F. Sottile 通过对组合 Hopf 代数范畴的研究, 证明了拟对称函数组合 Hopf 代数 ($\mathcal{QSym},\zeta_Q$) 与对称函数组合 Hopf 代数 ($\mathcal{Sym},\zeta_\mathcal{S}$) 分别是组合 Hopf 代数范畴与余交换的组合 Hopf 代数范畴的终对象.

曲阜师范大学数学科学学院的赵燕、高凤霞、李本星三位教授 2005 年先回顾了关于组合 Hopf 代数与图论中的基本概念及基础知识；并介绍了 3 个组合 Hopf 代数, 针对 3 个典型的分次连通 Hopf 代数结构, 将其分别赋予了一个非平凡的特征, 构造成了 3 个典型的组合 Hopf 代数: 完全图组合 Hopf 代数; Faà di Bruno 组合 Hopf 代数与具有全序顶点集的图的组合 Hopf 代数, 并分别讨论了它们的典型特征. 由于在组合 Hopf 代数范畴中, 拟对称函数组合 Hopf 代数 ($\mathcal{QSym},\zeta_Q$) 是终对象, 所以对于任何一个组合 Hopf 代数 ($\mathcal{H},\zeta$) 存在一个唯一的组合 Hopf 代数同态: $\zeta:(\mathcal{H},\zeta)\to(\mathcal{QSym},\zeta_Q)$. 他们还建立了例子中的组合 Hopf 代数与 ($\mathcal{QSym},\zeta_Q$) 之间的同态,并简单介绍了 Hilbert 级数且讨论了例子中的 Hilbert 级数的形式[①].

组合 Hopf 代数是二元组 $(\mathcal{H},\zeta)$,其中 $\mathcal{H} = \bigoplus_{n\geq 0}\mathcal{H}_n$ 是一个域 K 上的分次连通 Hopf 代数并且对任意 $n \geq 0$, $\dim(\mathcal{H}_n)$ 都是有限的. 这里 $\zeta:\mathcal{H} \to K$ 是一个特征,即保持乘法的线性函数,叫作 ζ 函数. 组合 Hopf 代数之间的同态 $\alpha:(\mathcal{H}',\zeta') \to (\mathcal{H},\zeta)$ 是分次 Hopf 代数同态,且满足 $\zeta' = \zeta \circ \alpha$. 域 K 上的组合 Hopf 代数连同它们之间的同态构成了域 K 上的组合 Hopf 代数范畴. 在组合 Hopf 代数中,除特征 ζ 外,还有 3 个典型的特征,分别命名为 Möbius 特征,Euler 特征与奇特征.

Möbius 特征:$\zeta^{-1} = \zeta \circ S_\mathcal{H}$,其中 $S_\mathcal{H}$ 是 $\mathcal{H}$ 的对极.

Euler 特征:$\chi = \bar{\zeta}\zeta$,其中 $\bar{\zeta}(h) = (-1)^n \zeta(h)$,$\forall h \in \mathcal{H}_n$.

奇特征:$\gamma = \bar{\zeta}^{-1}\zeta$.

在组合 Hopf 代数范畴中,M. Aguiar, N. Bergeron 和 F. Sottile 给出了如下重要结论:对任意的组合 Hopf 代数 $(\mathcal{H},\zeta)$,存在唯一的组合 Hopf 代数同态:$\Psi:(\mathcal{H},\zeta) \to (\mathcal{QSym},\zeta_Q)$,即拟对称函数组合 Hopf 代数 $(\mathcal{QSym},\zeta_Q)$ 是组合 Hopf 代数范畴的终对象.

设 G 是一个图,我们用 $V(G)$ 与 $E(G)$ 分别表示 G 的顶点集与边集.

简单图:不含平行边也不含环的图.

完全图:每一对不同的顶点均有一条边相连的简单

[①] 摘自《曲阜师范大学学报》,第 31 卷,第 4 期.

图.

空图:规定顶点集为空集的图称为空图,记为 ψ.

离散图:无边关联的点为孤立点,只含有孤立点的图为离散图.

两图的直和:指两图不相交的并. 记图 G 与图 H 的直和为 $G+H$.

导出子图:$U\subseteq V(G)$,则 $G|U$ 是导出子图,顶点集为 U,边集是 G 中两端点均在 U 中的边的集合.

设 G 是简单图,$E(G)$ 是其边集,S 是 $E(G)$ 的子集. $G|S$ 表示由 S 中的边及与 S 有关的点组成的图,称为 G 在边集 S 上的限制. $G \cdot S$ 表示在 G 中把 $E(G)-S$ 中的边去掉并把 S 在 G 中的所有端点合成一个点而得到的图,称为 G 到 S 的收缩.

如果在 $G \cdot (E(G)-S)$ 时没有环生成,则子集 $S \subseteq E(G)$ 称为边的闭子集. 如果一类简单图满足对于直和及到边的闭子集上的限制与收缩均封闭,则称这一类简单图为图的遗传族.

设 $G=\langle V,E \rangle$ 为一个无向图,若能将 V 分成 V_1 和 $V_2(V_1 \cup V_2 = V, V_1 \cap V_2 = \psi)$ 使得 G 中每条边的两个端点都是一个属于 V_1,另一个属于 V_2,则称 G 为二部图,记为 $G=\langle V_1,V_2,E \rangle$. 如图 1,都是二部图.

图 1

设 $\mathscr{G}$ 是一个图族,在导出子图与直和下保持封闭. 设

$\widetilde{\mathscr{G}}$ 是 $\mathscr{G}$ 中的图的同构类集合，$\widetilde{\mathscr{G}_0}$ 是 $\mathscr{G}$ 中连通图的同构类集合.

$\widetilde{\mathscr{G}}$ 中的乘积我们定义为 $[G] \cdot [H] = [G+H]$，$\forall [G],[H] \in \widetilde{\mathscr{G}}$，则 $\widetilde{\mathscr{G}}$ 可以看作是 $\widetilde{\mathscr{G}_0}$ 生成自由交换幺半群. 在 $\widetilde{\mathscr{G}}$ 上建立 Hopf 代数结构，其代数结构规定为多项式代数 $K[\widetilde{\mathscr{G}_0}]$

$$\Delta[G] = \sum_{U \subseteq V(G)} [G|U] \otimes [G|(V(G)-U)], \forall G \in \mathscr{G}$$

$$\varepsilon([G]) = \begin{cases} 1, G \text{ 为空图} \\ 0, \text{其他} \end{cases}$$

$$S[G] = \sum_{\pi} (-1)^{|\pi|} |\pi|! \pi \prod_{B \in \pi} [G|B]$$

其中 π 是取遍 $V(G)$ 的所有分拆.

特别地，设 $\mathscr{H}$ 是由 $\mathscr{G}$ 中所有完全图的不相交的并组成的集合. 我们用 x_n 表示含有 n 个顶点的完全图的同构类. 代数结构 $H(\mathscr{H})$ 为多项式代数 $K[x_1, x_2, \cdots]$，定义余代数结构如下

$$\Delta(x_n) = \sum_{k=0}^{n} \binom{n}{k} x_k \otimes x_{n-k}$$

$$\varepsilon(x_n) = \begin{cases} 1, n=0 \\ 0, \text{其他} \end{cases}$$

$$S(x_n) = \sum_{k=0}^{n} (-1)^k k! \, B_{n,k}(x_1, x_2, \cdots)$$

其中 $B_{n,k}(x_1, x_2, \cdots)$ 是部分 Bell 多项式

$$B_{n,k}(x_1, x_2, \cdots) = \sum_{\pi \in \prod_{n,k}} x_1^{\pi_1} x_2^{\pi_2} \cdots$$

其中 $\prod_{n,k}$ 是 $\{1, 2, \cdots, n\}$ 分成 k 部分的分拆的集合，π_i

是指对于每个分拆,分成的块的大小为 i 的块的个数. 上述域 $\mathcal{H}$ 上的 Hopf 代数结构我们记为 $H(\mathcal{H})$,它是一个交换余交换的 Hopf 代数. 设 $\deg(x_n) = n$, $H(\mathcal{H})$ 是分次连通的 Hopf 代数.

定义特征: $\zeta : H(\mathcal{H}) \to \dot{K}, x_n \mapsto \begin{cases} 1, n = 0 \text{ 或 } 1 \\ 0, \text{其他} \end{cases}$.

设 G 是一个连通的完全图,$\deg(x_n) = n$.

Möbius 特征

$$\zeta^{-1}(x_n) = \zeta \circ S(x_n)$$
$$= \sum_{k=0}^{n} (-1)^k k! \sum_{\pi \in \prod_{n,k}} \zeta(x_1^{\pi_1} x_2^{\pi_2} \cdots)$$
$$= (-1)^n n!$$

Euler 特征

$$\chi(x_n) = \bar{\zeta}\zeta(x_n) = \sum_{i=0}^{n} \binom{n}{i} \bar{\zeta}(x_i) \zeta(x_{n-i})$$
$$= \begin{cases} 1, n = 0 \\ (-1)^1 \cdot 1 + (-1)^0 \cdot 1 = 0, n = 1 \\ 2(-1)^1 \cdot 1 = -2, n = 2 \\ 0, n > 2 \end{cases}$$

奇特征

$$\gamma(x_n) = \bar{\zeta}^{-1}\zeta(x_n) = \sum_{i=0}^{n} \binom{n}{i} \bar{\zeta}^{-1}(x_i) \zeta(x_{n-i})$$
$$= \binom{n}{0} \bar{\zeta}^{-1}(x_0) \zeta(x_n) +$$
$$\binom{n}{1} \bar{\zeta}^{-1}(x_1) \zeta(x_{n-1}) + \cdots + \binom{n}{n} \bar{\zeta}^{-1}(x_n) \zeta(x_0)$$
$$= \binom{n}{n-1} \bar{\zeta}^{-1}(x_{n-1}) \zeta(x_1) + \binom{n}{n} \bar{\zeta}^{-1}(x_n) \zeta(x_0)$$

$$= n(-1)^{n-1}(-1)^{n-1}(n-1)! +$$
$$(-1)^n(-1)^n n!$$
$$= 2 \times n!, n \neq 0$$
$$\gamma(x_n) = 1, n = 0$$

设 $\mathscr{G}$ 是一个图的遗传族,定义图 $\mathscr{G}$ 与 H 弱同构 $\Leftrightarrow \mathscr{G}$ 与 H 删掉所有孤立点后同构. $[G]$ 表示图 $\mathscr{G}$ 所在的弱同构类. $\widetilde{\mathscr{G}}$ 表示 $\mathscr{G}$ 的弱同构类的集合. $H(\mathscr{G})$ 表示 $\mathscr{G}$ 在 K 上的幺半群代数并定义下列运算:

余乘法: $[G] = \sum_{S \in \mathscr{L}(G)} [G|S] \otimes [G \cdot (E(G) - S)]$,

其中 $\mathscr{L}(G)$ 是图 G 的收缩格,即所有 $E(G)$ 的闭子集组成的格,以包含它的偏序. 所谓 $S \in \mathscr{L}(G)$ 是闭子集,指在 $G \cdot (E(G) - S)$ 中不会有环出现. (由于限制与收缩是交换的过程,即如果 $S_1 + S_2 + S_3 = E(G)$,有 $(G|(S_1 + S_2)) \cdot S_1 = (G|(S_1 + S_3)) \cdot S_1$,那么余乘法满足余结合性.)

余单位: $\varepsilon[G] = \begin{cases} 1, 若 E(G) = \psi \\ 0, 其他 \end{cases}$,可以得到 $H(\mathscr{G})$ 是交换非余交换的 Hopf 代数.

特别地,设 $\mathscr{H}$ 是由完全图的不相交的并组成的遗传族. 用 x_n 表示 $n+1$ 个顶点的完全图的同构类. 图的类型为 x_n 的收缩格为 $n+1$ 个元素集合的全分拆格.

令 $H(\mathscr{H})$ 的代数结构同构于 $K[x_1, x_2, \cdots]$,则

$$\Delta(x_n) = \sum_{k=0}^{n} B_{n+1,k+1}(1, x_1, x_2, \cdots) \otimes x_k$$

$$\varepsilon(x_n) = \begin{cases} 1, 若 n = 0 \\ 0, 其他 \end{cases}$$

$$S(x_n) = \sum_{k \geq 1} (-1)^k B_{n+k,k}(0, x_1, x_2, \cdots)$$

$$= \sum_{k \geqslant 1} (-1)^k \sum_{\pi \in \prod_{n+k,k}} 0^{\pi_1} x_1^{\pi_2} x_2^{\pi_3} \cdots$$

按边的条数进行分次，这是一个交换非余交换的分次连通的 Hopf 代数，记为 $\mathscr{F} = \mathscr{H}(\mathscr{H})$，称为 Faà di Bruno Hopf 代数. 因为分次是按边的条数，所以，$\deg(x_1) = 1$，$\deg(x_2) = 3, \deg(x_3) = 6, \cdots$

定义特征：$\zeta : \mathscr{F} \to K, x_n \mapsto \begin{cases} 1, n = 0, 1 \\ 0, 其他 \end{cases}$.

Möbius 特征

$$\zeta^{-1}(x_n) = \zeta \circ S(x_n) = \sum_{k \geqslant 1} (-1)^k \sum_{\pi \in \prod_{n+k,k}} \zeta(0^{\pi_1} x_1^{\pi_2} x_2^{\pi_3} \cdots)$$

$$= (-1)^n \frac{C_{2n}^2 C_{2n-2}^2 \cdots C_2^2}{n!} [\zeta(x_1)]^n$$

$$= (-1)^n \frac{C_{2n}^2 C_{2n-2}^2 \cdots C_2^2}{n!}$$

$$= (-1)^n \frac{(2n)!}{n! \, 2^n}$$

Euler 特征

$$\chi(x_n) = \bar{\zeta}\zeta(x_n) = \sum_{k=0}^{n} \sum_{\pi \in \prod_{n+1,k+1}} \bar{\zeta}(1^{\pi_1} x_1^{\pi_2} x_2^{\pi_3} \cdots) \zeta(x_k)$$

$$= \sum_{\pi \in \prod_{n+1,1}} \bar{\zeta}(1^{\pi_1} x_1^{\pi_2} x_2^{\pi_3} \cdots) \zeta(x_0) +$$

$$\sum_{\pi \in \prod_{n+1,2}} \bar{\zeta}(1^{\pi_1} x_1^{\pi_2} x_2^{\pi_3} \cdots) \zeta(x_1)$$

$$= \begin{cases} 1, n=0 \\ \bar{\zeta}(x_1^1) + \bar{\zeta}(1^2) = (-1)^1 + (-1)^0 = 0, n=1 \\ 3\bar{\zeta}(1^1 x_1^1) = 3((-1)^0(-1)^1) = -3, n=2 \\ 3\bar{\zeta}(x_1^2) = 3((-1)^1 \cdot 1)^2 = 3, n=3 \\ 0, \text{其他} \end{cases}$$

奇特征的情况比较复杂

$$\gamma(x_n) = \bar{\zeta}^{-1}\zeta(x_n) = \sum_{k=0}^{n} \sum_{\pi \in \prod_{n+1,k+1}} \bar{\zeta}^{-1}(1^{\pi_1} x_1^{\pi_2} x_2^{\pi_3} \cdots) \zeta(x_k)$$

$$= \sum_{\pi \in \prod_{n+1,1}} \bar{\zeta}^{-1}(1^{\pi_1} x_1^{\pi_2} x_2^{\pi_3} \cdots) \zeta(x_0) +$$

$$\sum_{\pi \in \prod_{n+1,2}} \bar{\zeta}^{-1}(1^{\pi_1} x_1^{\pi_2} x_2^{\pi_3} \cdots) \zeta(x_1)$$

分为以下几种情况进行讨论：

1. $n=0$ 时，$\gamma(x_n) = 1$；

2. $n>0$ 且 n 为奇数时

$$\gamma(x_n) = \bar{\zeta}^{-1}(x_n) + C_{n+1}^1 \bar{\zeta}^{-1}(x_{n-1}) +$$
$$C_{n+1}^2 \bar{\zeta}^{-1}(x_1) \bar{\zeta}^{-1}(x_{n-2}) + \cdots +$$
$$C_{n+1}^{\frac{n-1}{2}} \bar{\zeta}^{-1}(x_{\frac{n-1}{2}-1}) \bar{\zeta}^{-1}(x_{n-\frac{n-1}{2}}) +$$
$$\frac{1}{2} C_{n+1}^{\frac{n+1}{2}} [\bar{\zeta}^{-1}(x_{\frac{n+1}{2}-1})]^2$$

$$= \bar{\zeta}^{-1}(x_n) + \sum_{i=1}^{\frac{n-1}{2}} C_{n+1}^i \bar{\zeta}^{-1}(x_{i-1}) \bar{\zeta}^{-1}(x_{n-i}) +$$
$$\frac{1}{2} C_{n+1}^{\frac{n+1}{2}} [\bar{\zeta}^{-1}(x_{\frac{n+1}{2}-1})]^2$$

$$= \frac{(2n)!}{n! \, 2^n} + \sum_{i=1}^{\frac{n-1}{2}} C_{n+1}^i \frac{[2(i-1)]!}{(i-1)!} \frac{[2(n-i)]!}{(n-i)! \, 2^{n-1}} +$$

$$\frac{1}{2}C_{n+1}^{\frac{n+1}{2}}\left(\frac{(n-1)!}{\left(\frac{n-1}{2}\right)!\ 2^{\frac{n-1}{2}}}\right)^2$$

3. $n>0$ 且 n 为偶数时

$$\begin{aligned}\gamma(x_n) &= \bar{\zeta}^{-1}(x_n)+C_{n+1}^{1}\bar{\zeta}^{-1}(x_{n-1})+C_{n+1}^{2}\bar{\zeta}^{-1}(x_1)\bar{\zeta}^{-1}(x_{n-2})+\cdots+\\ &\quad C_{n+1}^{\frac{n}{2}}\bar{\zeta}^{-1}(x_{\frac{n}{2}-1})\bar{\zeta}^{-1}(x_{\frac{n}{2}})\\ &= \bar{\zeta}^{-1}(x_n)+\sum_{i=1}^{\frac{n}{2}}C_{n+1}^{i}\bar{\zeta}^{-1}(x_{i-1})\bar{\zeta}^{-1}(x_{n-i})\\ &= \frac{(2n)!}{n!\ 2^n}+\sum_{i=1}^{\frac{n}{2}}C_{n+1}^{i}\frac{[2(i-1)]!\ [2(n-i)]!}{(i-1)!\ (n-i)!\ 2^{n-1}}\end{aligned}$$

设 $\mathscr{G}$ 是一族具有全序顶点集的简单图，对于不相交的并集与导出子图是封闭的. 在 $\mathscr{G}$ 上定义同构类：$G_1, G_2\in\mathscr{G}$，则 $G_1\cong G_2\Leftrightarrow$ 存在保持序的图同构 $f: V(G_1)\rightarrow V(G_2)$. 用 $\langle G\rangle$ 表示 $G\in\mathscr{G}$ 所在的同构类，表示 $\widetilde{\mathscr{G}}$ 中 $\mathscr{G}$ 所有的同构类的集合. 显然 $\widetilde{\mathscr{G}}$ 是一个含幺半群，乘法是图的直和. 因为图 $G\in\mathscr{G}$ 是全序顶点集，如图 2，所以显然乘法是非交换的. 用 $H_l(\mathscr{G})$ 表示 $\widetilde{\mathscr{G}}$ 在 K 上的幺半群代数，定义下列运算：

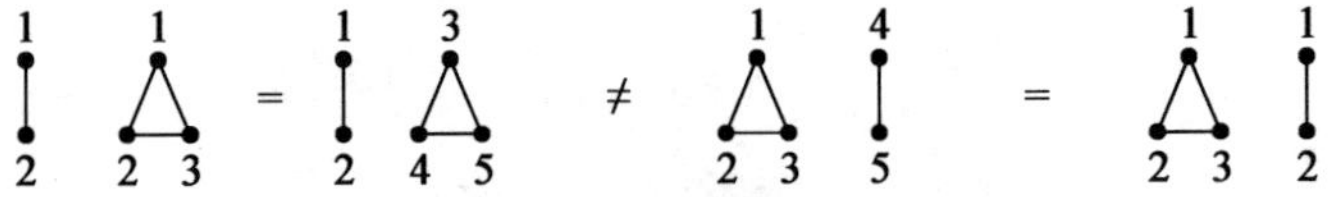

图 2

余乘法：$\triangle\langle G\rangle=\sum_{U\subseteq V(G)}\langle G|U\rangle\otimes\langle G|(V(G)-U)\rangle$；

余单位：$\varepsilon\langle G\rangle=\begin{cases}1, 若\ V(G)=\phi\\ 0, 其他\end{cases}$；

对极:$S\langle G\rangle = \sum_{\pi}(-1)^{|\pi|}|\pi|!\prod_{B\in\pi}\langle G|B\rangle$,$\pi$是$V(G)$的分拆,可见$H_l(\mathcal{G})$是一个非交换、余交换的分次连通的 Hopf 代数,$\deg\langle G\rangle = |V(G)|$.

定义特征:$\zeta:H_l(\mathcal{G})\to K,\langle G\rangle\mapsto\begin{cases}1,G\text{ 为空图或离散图}\\0,\text{其他}\end{cases}$.

可见,ζ是保持乘法的线性函数,$(H_l(\mathcal{G}),\zeta)$是组合 Hopf 代数. 若$\langle G\rangle\in H_l(\mathcal{G})_{(n)}$,不妨设 G 为连通图时的情况,则:

Möbius 特征

$$\zeta^{-1}(\langle G\rangle) = \zeta\circ S(\langle G\rangle) = (-1)^n n!$$

Euler 特征

$$\chi(\langle G\rangle) = \bar{\zeta}\zeta(\langle G\rangle) = \sum_{U\in V(G)}\bar{\zeta}(\langle G|U\rangle)\zeta(\langle G|(V(G)-U)\rangle)$$

$$=\begin{cases}1,n=0\\(-1)^m+(-1)^{n-m},n\geqslant 1,G=\langle V_1,V_2,E\rangle\text{ 为二部图},|V_1|=m\\0,\text{其他}\end{cases}$$

奇特征

$$\gamma(\langle G\rangle) = \bar{\zeta}^{-1}\zeta(\langle G\rangle) = \sum_{G\in V(G)}\bar{\zeta}^{-1}(\langle G|S\rangle)\zeta(\langle G|(V(G)-U)\rangle)$$

$$=\begin{cases}1,n=0\\m!+(n-m)!,n\geqslant 1,\text{且 }G=\langle V_1,V_2,E\rangle\text{ 为二部图},|V_1|=m\\0,\text{其他}\end{cases}$$

我们知道$(\mathcal{QSym},\zeta_Q)$是组合 Hopf 代数范畴的终对象,则以上例子中组合 Hopf 代数到$(\mathcal{QSym},\zeta_Q)$一定分别存在一个唯一的组合 Hopf 代数同态. 我们以全序顶点集的简单图族上的组合 Hopf 代数为例,这里只考虑 G 为连通图的情况. 若

$$\Psi:(H_l(\mathscr{G}),\zeta)\to(\mathscr{QSym},\zeta_Q),\forall\langle G\rangle\in H_l(\mathscr{G})_{(n)}$$

则 $\Psi\langle G\rangle = \sum_{\alpha=n}\zeta_\alpha(\langle G\rangle)M_\alpha, \alpha=(a_1,\cdots,a_k)$,其中

$$\zeta_\alpha:H_l(\mathscr{G})\xrightarrow{\Delta^{(k-1)}}H_l(\mathscr{G})^{\otimes k}\longrightarrow H_l(\mathscr{G})_{a_1}\otimes\cdots\otimes_1(\mathscr{G})_{(a_k)}\xrightarrow{\zeta^{\otimes k}}K$$

$$\zeta_\alpha\langle G\rangle = \mu\zeta^{\otimes k}\left(\binom{n}{1}\langle G_1\rangle\otimes\cdots\otimes\binom{1}{1}\langle G_1\rangle\right)=n!$$

其中 G_1 为单点图. 所以, $\Psi\langle G\rangle = n! \cdot M_{\underbrace{(1,1,\cdots,1)}_{n\uparrow}}$. 类似可得其他例子到 $(\mathscr{QSym},\zeta_Q)$ 的组合 Hopf 代数同态.

定义 若 $V=\bigoplus_{n\geq 0}V_n$ 是一个分次的向量空间,且每个 V_n 是有限维的,则 $V(x)=\sum_{n\geq 0}\dim V_n x^n$ 称为 V 的 Hilbert 级数.

下面我们分别讨论以上 3 个例子中的 Hilbert 级数.

1. 对于完全图组合 Hopf 代数 $(H(\mathscr{H}),\zeta)$, $\dim H(\mathscr{H})_{(n)}$ 为含有 n 个顶点的完全图生成的集合的维数. 设 a_i 表示共含 i 个顶点的完全图生成的图集的同构类的个数,则 $\dim H(\mathscr{H})_{(n)}$ 为 n 元一次方程 $a_1+2a_2+\cdots+na_n=n$ 正整数解的个数,由数学归纳法可知解的个数为 n,所以 $(H(\mathscr{H}),\zeta)$ 的 Hilbert 级数为

$$H(\mathscr{H})(x)=1+\sum_{n\geq 1}nx^n$$

2. 对于 Faà di Bruno Hopf 代数 $\mathscr{F}$,按边的条数进行分次. 我们首先考虑最简单的几个分次空间的维数. 显然,有 $\dim(\mathscr{F}_{(1)})=1, \dim(\mathscr{F}_{(2)})=1$,对于 $\dim(\mathscr{F}_{(3)})$ 因为含有 3 条边的情况只有两种 ⋀ 和 ∣∣∣,所以 $\dim(\mathscr{F}_{(3)})=2$. 而含有 4 条边的情况也有两种 ⋃ 和 ∣∣∣∣,所以 $\dim(\mathscr{F}_{(4)})=2$. 对于较复杂的情况,我们将 $\dim(\mathscr{F}_{(n)})$ 归

纳为 $k-1$ 元一次方程 $y_1 + C_3^2 y_3 + \cdots + C_k^2 y_k = n$(其中 k 满足 $C_k^2 \leq n < C_{k+1}^2$)的正整数解的个数,我们记为 A_n. 其中 C_k^2 是含有 k 个顶点的完全图边的条数. 所以 Hilbert 级数为

$$\mathscr{A}(x) = \sum_{n \geq 0} A_n x^n$$

3. 对于全序顶点集的组合 Hopf 代数 $H_l(\mathscr{G})$, n 次齐次分量的维数,即为 n 个全序顶点能生成图的同构类的个数. n 个全序顶点能生成 $\binom{n}{2}$ 条边,所以共有 $2^{\binom{n}{2}}$ 种不同构的图类. $\dim \mathscr{H}_l(\mathscr{G})_{(n)} = 2^{\binom{n}{2}}$. 其 Hilbert 级数为

$$\mathscr{H}_l(\mathscr{G})(x) = 1 + x^1 + \sum_{n \geq 2} 2^{\binom{n}{2}} x^n$$

曲阜师范大学数学科学学院的程腾、王顶国、程诚三位教授 2014 年把 Run-qiang Jian 文中的 H 为 Hopf 代数的情况推广到 H 为 Hopf(余)拟群,其主要结论:设 H 是 Hopf 拟群, (M, φ) 是一右拟 H-Hopf 模代数,则 (M, P) 是权为 -1 的 Rota-Baxter 代数[①].

Hopf 代数的一般化有相当长的历史, Drinfeld G. V. 在 *Quasi-Hopf algebras* 中弱化了(余)结合性,定义了拟 Hopf 代数. Van Daele 在 *Multilier Hopf algebras* 中弱化了单位,定义了乘子 Hopf 代数. Bohm, Nill 和 Szlachanyi 在 *Weak Hopf algebras* I 中定义了弱 Hopf 代数,既不要求单位满足余结合律和余单位不满足结合律. Klim J. 和 Majid S. 在 *Hopf quasigroups and the algebras T-sphere* 中介

---

① 摘自《曲阜师范大学学报》,第 40 卷,第 3 期.

绍了 Hopf(余)拟群. Brzezinski 在 *Hopf modules and fundamental theorem for Hopf(co) quasigroups* 中定义了 Hopf (余)拟群(quasigroups)上的 Hopf 模,并证明了这样一个事实:对任意的 Hopf(余)拟群 H,左、右 H-Hopf 模范畴等价于向量空间范畴. 本文主要是将 *Construction of Rota-Baxter algebras via Hopf module algebras* 中 H 为 Hopf 代数的情况推广到 H 为 Hopf(余)拟群,并得到一系列结果. 其主要结论:设 H 是 Hopf 拟群,(M,φ) 是一右拟 H-Hopf 模代数,则 (M,P) 是权为 -1 的 Rota-Baxter 代数.

定义 1[①] 设 H 是一向量空间,且 H 是一有单位的代数,有余单位的余代数,其乘积为 $\mu:H\otimes H\to H$,单位为 $\eta:K\to H$,余积为 $\Delta:H\to H\otimes H$,余单位为 $\varepsilon:H\to K$ 且 Δ,ε 是代数同态.

称 H 为 Hopf 拟群,如果 Δ 是余结合的且存在线性映射 $S:H\to H$,使得

$$\mu(I\otimes\mu)(S\otimes I\otimes I)(\Delta\otimes I) = \varepsilon\otimes I$$
$$= \mu(\mu\otimes I)(I\otimes I\otimes S)(\Delta\otimes I)$$
$$\mu(\mu\otimes I)(I\otimes I\otimes S)(I\otimes\Delta) = I\otimes\varepsilon$$
$$= \mu(\mu\otimes I)(I\otimes S\otimes I)(I\otimes\Delta)$$

称 H 为 Hopf 余拟群,如果 μ 是结合的且存在线性映射 $S:H\to H$,使得

$$(\mu\otimes I)(S\otimes I\otimes I)(I\otimes\Delta)\Delta = \eta\otimes I$$
$$= (\mu\otimes I)(I\otimes S\otimes I)(I\otimes\Delta)\Delta$$
$$(I\otimes\mu)(I\otimes I\otimes S)(\Delta\otimes I)\Delta = I\otimes\eta$$

[①] Klim J, Majid S. Hopf quasigroups and the algebras 7-sphere. Algebra,2010,323: 3067-3110.

$$= (I\otimes\mu)(I\otimes S\otimes I)(\Delta\otimes I)\Delta$$

定义 2[①] 设 H 是一 Hopf 拟群,M 是一有余结合律和余单位的右 H 余模,其余作用记为 ρ^M,且 M 是一有单位的右 H 模,其作用记为 ρ_M,称 M 为右 H-Hopf 模,若

$$\rho_M(\rho_M\otimes I)(I\otimes I\otimes S)(I\otimes\Delta) = I\otimes\varepsilon$$
$$= \rho_M(\rho_M\otimes I)(I\otimes S\otimes I)(I\otimes\Delta)$$

(1)

$$\rho^M \circ \rho_M = (\rho_M\otimes\mu)(I\otimes\tau\otimes I)(\rho^M\otimes\Delta) \quad (2)$$

引理 3 设 H 是 Hopf 拟群,且 M 是右 H-Hopf 模,则对任意的 $m\in M, h\in H$ 有

$$\rho^M(m_{(0)}S(m_{(1)})) = m_{(0)}\cdot S(m_{(1)})\otimes 1_H$$
$$\rho^M((m_{(0)}\cdot S(m_{(1)}))\cdot h) = (m_{(0)}\cdot S(m_{(1)}))\cdot h_{(1)}\otimes h_{(2)}$$

证明

$$\rho^M(m_{(0)}\cdot S(m_{(1)})) = (m_{(0)}\cdot S(m_{(1)}))_{(0)}\otimes(m_{(0)}\cdot S(m_{(1)}))_{(1)}$$
$$\stackrel{(2)}{=} m_{(0)(0)}\cdot S(m_{(1)})_{(1)}\otimes m_{(0)(1)}\cdot S(m_{(1)})_{(2)}$$
$$= m_{(0)}\cdot S(m_{(3)})\otimes m_{(1)}S(m_{(2)})$$
$$= m_{(0)}\cdot S(\varepsilon(m_{(1)}m_{(2)}))\otimes 1_H$$
$$= m_{(0)}\cdot S(m_{(1)})\otimes 1_H$$

$$\rho^M((m_{(0)}\cdot S(m_{(1)}))\cdot h)_{(0)}\otimes((m_{(0)}\cdot S(m_{(1)}))\cdot h)_{(1)}$$
$$= (m_{(0)}\cdot S(m_{(1)}))_{(0)}\cdot h_{(1)}\otimes(m_{(0)}\cdot S(m_{(1)}))_{(1)}h_{(2)}$$
$$\stackrel{(3)}{=} (m_{(0)}\cdot S(m_{(1)}))h_{(1)}\otimes h_{(2)} \quad (3)$$

[①] Tomasz Brzezinski. Hopf modules and the fundamental theorem for Hopf(co)quasi-groups. International Electronic Journal of Algebra vol,2010,8:114-128.

定义 4[①] 设 H 是 Hopf 余拟群，H 是一有余单位的右 H-余模，其余作用记为 ρ^M，且 M 是一具有结合律和单位的右 H-模，其作用记为 ρ_M，称 M 为右 H-Hopf 模，若

$$(I\otimes\mu)(I\otimes I\otimes S)(\rho_M\otimes I)\rho^M = I\otimes\eta$$
$$= (I\otimes\mu)(I\otimes S\otimes I)(\rho_M\otimes I)\rho^M$$
$$\rho^M \circ \rho_M = (\rho_M\otimes\mu)(I\otimes\tau\otimes I)(\rho^M\otimes\Delta)$$

定义 5 设 H 是 Hopf 拟群，右拟 H-Hopf 模代数是一右 H-Hopf 模 M，且 M 有一结合乘法 $\varphi:M\otimes M\to M$，使得对任意 $h\in H$ 和 $m,n\in M$，有

$$(mn)\cdot h = m(n\cdot h), \rho^M(mn) = m_{(0)}n_{(0)}\otimes m_{(1)}n_{(1)}$$

其中 $mn=\varphi(m\otimes n)$。

引理 6 定义 $P:M\to M, P(m)=m_{(0)}\cdot S(m_{(1)})$，则有 $P^2=P$，且 P 是 M 到 M^H 的投射，其中 $M^H=\{m\in M|\rho^M(m)=m\otimes 1_H\}$。

证明 $P^2(m)=P(m_{(0)}\cdot S(m_{(1)}))=\rho^M(I\otimes S)\cdot((m_{(0)}\cdot S(m_{(1)}))\otimes 1_H)=m_{(0)}\cdot S(m_{(1)})$，即有 $P^2=P$。
由引理 3 得

$$\rho^M((m_{(0)}\cdot S(m_{(1)}))\cdot h)$$
$$=((m_{(0)}\cdot S(m_{(1)}))\cdot h)_{(0)}\otimes((m_{(0)}\cdot S(m_{(1)}))\cdot h)_{(1)}$$

定义 7 设 K 为域，$\lambda\in K$，称有序对 (R,P) 为一个权为 λ 的 Rota-Baxter 代数，若 R 是代数，P 是 R 的线性自同态，且满足

$$P(x)P(y)=P(xp(y))+P(P(x)y)+\lambda P(xy)$$

称 P 为 Rota-Baxter 算子。

[①] Tomasz Brzezinski. Hopf modules and the fundamental theorem for Hopf(co)quasi-groups. International Electronic Journal of Algebra vol,2010,8:114-128.

定理 8 设 H 是 Hopf 拟群,(M,φ) 是一右拟 H-Hopf 模代数,则 (M,P) 是权为 -1 的 Rota-Baxter 代数(P 为引理 6 中的定义).

证明 对任意 $m,n \in M$,有

$P(mP(n)) + P(P(m)n) - P(mn)$
$= (mP(n))_{(0)} \cdot S((mP(n))_{(1)}) + (P(m)n)_{(0)} \cdot S((P(m)n)_{(1)}) - (mn)_{(0)} \cdot S((mn)_{(1)})$
$= m_{(0)}P(n)_{(0)} \cdot S(m_{(1)}P(n)_{(1)}) + (P(m)_{(0)}n_{(0)}) \cdot S(P(m)_{(1)}n_{(1)}) - (m_{(0)}n_{(0)}) \cdot S(m_{(1)}n_{(1)})$
$= (m_{(0)}P(n)) \cdot S(m_{(1)}) + P(m)n_{(0)} \cdot S(n_{(1)}) - (m_{(0)}n_{(0)}) \cdot S(m_{(1)}n_{(1)})$
$= P(m)(n_{(0)} \cdot S(n_{(1)}) + (m_{(0)}P(n)) \cdot S(m_{(1)}) - m_{(0)}(n_{(0)} \cdot S(n_{(1)})) \cdot S(m_{(1)})$
$= P(m)P(n)$

定义 9 设 (M,φ) 是一右拟 H-Hopf 模代数,定义 M 与 H 的扭碎积以 $M \otimes H$ 为底空间,运算为

$$(m\#h)(n\#k) = m(h_{(1)} \cdot n)\#h_{(2)}k$$

其中 $m,n \in M;h,k \in H.$

定理 10 设 (M,φ) 是一右拟 H-Hopf 模代数,μ 满足结合律,其中 H 为 Hopf 拟群,则有 $M\#H$ 是右拟 H-Hopf 模代数,其作用如下

$$(m\#h)k = m\#hk, \rho^{M\#H}(m\#h) = (m\#h_{(1)}) \otimes h_{(2)}$$

其中 $m,n \in M;h,k \in H.$

因此有 $P(m\#h) = m\#\varepsilon(h)1_H$ 是权为 -1 的幂等的 Rota-Baxter 代数.

证明 首先证 $M\#H$ 是右 H-Hopf 模,由条件可知 $M\#H$ 是一有余结合律和余单位的右 H 余模,且 $M\#H$ 是一有单

位的右 H 模,下证模的相容条件: $\forall : m, n \in M, h, k \in H$,有
$$\rho_{M\#H}(\rho_{M\#H}\otimes I)(I\otimes I\otimes S)(I\otimes\Delta)(m\#h\otimes k)$$
$$=\rho_{M\#H}(\rho_{M\#H}\otimes I)(m\#h\otimes k_{(1)}\otimes S(k_{(2)}))$$
$$=(m\#h)k_{(1)}S(k_{(2)})=(I\otimes\varepsilon)(m\#h\otimes k)$$

即式(1)左边成立. 同理有式(1)右边成立
$$\rho^{M\#H}\circ\rho_{M\#H}(m\#h\otimes k)=\rho^{M\#H}(m\#hk)$$
$$=m\#(hk)_{(1)}\otimes(mh)_{(2)}$$
$$(\rho_{M\#H}\otimes\mu)(I\otimes\tau\otimes I)(\rho^{M\#H}\otimes\Delta)(m\#h\otimes k)$$
$$=(\rho_{M\#H}\otimes\mu)(I\otimes\tau\otimes I)(m\#h_{(1)}\otimes h_{(2)}\otimes k_{(1)}\otimes k_{(2)})$$
$$=m\#h_{(1)}k_{(1)}\otimes h_{(2)}k_{(2)}$$

因 H 是 Hopf 拟群,即有 Δ 是代数同态,故有式(2)成立.

最后证构成右拟 H-Hopf 模代数的相容条件.

对 $\forall : m, n \in M, h, k \in H, l \in H$,有
$$((m\#h)(n\#k))l=(m(h_{(1)}n)\#h_{(2)}k)l$$
$$=(m(h_{(1)}n)\#h_{(2)}k)l$$
$$=m(h_{(1)}n)\#h_{(2)}kl$$
$$=(m\#h)((n\#k)l)$$
$$\rho^{M\#H}((m\#h)(n\#k))=\rho^{M\#H}(m(h_{(1)}n)\#h_{(2)}k)$$
$$=(m(h_{(1)}n)\#h_{(2)}k_{(1)})\otimes h_{(3)}k_{(2)}$$
$$=(m\#h_{(1)})(n\#k_{(1)})\otimes h_{(2)}k_{(2)}$$
$$=(m\#h_{(1)}\otimes h_{(2)})((n\#k_{(1)})\otimes k_{(2)})$$
$$=(m\#h)_{(0)}(n\#h)_{(0)}\otimes(m\#h)_{(1)}(n\#h)_{(1)}$$

引理 11 设 H 为 Hopf 余拟群,且 M 是 H-Hopf 模. 对任意 $m \in M$,有
$$\rho^M(m_{(0)}\cdot S(m_{(1)}))=m_{(0)}\cdot S(m_{(1)})\otimes 1_H$$

$\rho^M((m_{(0)} \cdot S(m_{(1)}))h) = (m_{(0)}S(m_{(1)})) \cdot h_{(1)} \otimes h_{(2)}$

证明 类似于引理 3.

定义 12 设 H 为 Hopf 余拟群,定义余拟 H-Hopf 模代数是一右 H-Hopf 模 M,且 M 有一结合乘法 $\varphi: M \otimes M \to M$,使对任意 $h \in H, m, n \in M$,有

$(mn) \cdot h = m(n \cdot h), \rho^M(mn) = m_{(0)}n_{(0)} \otimes m_{(1)}n_{(1)}$

其中 $mn = \varphi(mn)$.

引理 13 定义 $P: M \to M, P(m) = m_{(0)} \cdot S(m_{(1)})$,则有 $P^2 = P$,且 P 是 M 到 M^H 的投射,其中 $M^H = \{m \in M | \rho^M = m \otimes 1_H\}$.

证明 类似于引理 6 的证明.

定理 14 设 H 是 Hopf 余拟群,(M, φ) 是一右余拟 H-Hopf模代数,则 (M, P) 是权为 -1 的 Rota-Baxter 代数,其中 P 为引理 13 中定义的.

证明 类似于定理 8.

定理 15 设 (M, φ) 是一右余拟 H-Hopf 模代数,其中 H 为 Hopf 余拟群,ε 满足余结合律,则有 $M\#H$ 右余拟 H-Hopf模代数,其作用如下,对 $\forall: k, h \in H, m \in M$,有

$(m\#h) \cdot k = m\#hk, \rho^{M\#H}(m\#h) = (m\#h_{(1)}) \otimes h_{(2)}$

因此有 $P(m\#h) = m\#\varepsilon(h)1_H$ 是权为 -1 的幂等的 Rota-Baxter 算子.

证明

$(I \otimes \mu)(I \otimes I \otimes S)(\rho^{M\#H} \otimes I)\rho^{M\#H}(m\#H)$
$= (I \otimes \mu)(I \otimes I \otimes S)(m\#h_{(1)} \otimes h_{(2)} \otimes h_{(3)})$
$= m\#h_{(1)} \otimes h_{(2)}S(h_{(3)})$
$= m\#h_{(1)} \otimes \varepsilon(h_{(2)}) = m\#h$

因为 ε 是余结合的. 其余部分类似于定理 10 的证明.

青海民族大学数学与统计学院的周淑云教授 2013 年介绍了 Rota-Baxter 代数的基本概念, 并介绍了 Rota-Baxter 代数的许多例子, 从这些例子中我们可以看到 Rota 构造自由 Rota-Baxter 代数的思想, 自由 Rota-Baxter 代数是 Rota-Baxter 代数研究的基础. 她还介绍了 Rota-Baxter 代数的基本性质[①].

1960 年, G. Baxter 发表了关于 Rota-Baxter 代数(也称 Baxter 代数)的一篇重要论文[②], 将概率论中的独立随机变量部分和过程的振荡性质的一个基本恒等式——Spitzer 恒等式和其他重要的恒等式进行了推演, 随即这些恒等式立刻引起了许多著名的数学家的注意, 如 F. V. Atkinson, P. Cartier 和 G. - C. Rota, 他们对此做了进一步的研究, 并取得了许多好的结果. 特别是 Rota 注意到了这些恒等式在代数学和组合数学中的重要作用, 并从 20 世纪 60 年代后期到 90 年代将 Baxter 算子和组合恒等式联系在一起[③④], 且他还提出了今后研究工作的方向[⑤]. Rota 在

---

① 摘自《青海师范大学学报》(自然科学版), 2013 年, 第 2 期.

② G. Baxter, An analytic problem whose solution follows from a simple algebraic identity. Pacific J. Math. , 10, (1960) :731-742.

③ G. -C. Rota, Baxter algebras and combinatorial identities Ⅱ. Bull. Amer. Math. Soc. , 75(1969) , :325-329.

④ G. -C. Rota, Baxter algebras and combinatorial identities Ⅱ. Bull. Amer. Math. Soc. 75(1969) :330-334.

⑤ G. -C. Rota, Ten mathematics problems I will never solve. Invied address at the joint meeting of the American Mathematical Society and the Mexican Mathematical Society, Oaxaca. Mexico, December 6(1997). DMV Mittellungen Heft 2(1998) :45-52.

此集合上构造了自由 Baxter 代数的一个显示结构,并由这个结构结合 Waring 公式给出了 Spitzer 恒等式的一个证明,并选取了合适的 Baxter 代数对称群的表示理论进行了讨论,还应用 Baxter 算子简化对称函数和 Möbius 反演的一些计算. Cartier[1] 于 20 世纪 70 年代也构造了自由 Baxter 代数的一个显示结构,弄清楚了 Baxter 代数和对称多项式之间的联系. Baxter 代数还被广泛应用于代数学、几何学、Lie 代数等多方面的研究,如累次积分、微分代数、对称函数、超几何函数和量子群,等等.

1986 年, V. G. Drinfeld 在 Berkeley 国际数学家大会上报告了《量子群》[2]一文,引起了数学界的广泛关注,他把统计学和 Yang-Baxter(以著名的物理学家 C. N. Yang 和 R. Baxter 命名)方程的波动力学问题的研究转化为 Hopf 代数的研究,有其深刻的物理背景. 量子群是 Lie 群、Lie 代数与代数发展到一定阶段的产物,综合了物理学与数学的许多分支的思想和内容,具有十分广泛的理论意义和应用范围. 在代数方面,它将群与代数的表示理论有机地结合起来. 自 20 世纪 80 年代以来,一些数学家在研究 Yang-Baxter 方程时,在 Lie 代数中发现了 Rota-Baxter 恒等式,从而引起了很多数学家和物理学家的兴趣. 1998 年, Winkel[3] 在研究 Baxter 序列方面的工作之后, Connes

---

[1] P. Cartier, On the structure of free Baxter algebras. Adv. Math. ,9(1972):253 – 265.

[2] V. G. Drinfeld, "Quantum groups" Proceedings of the International Congress of Mathematicians, Vol. 1, 2 (Berkeley, Calif. , 1986) , 798-820, Amer. Math. Soc. , Providence, RI, 1987.

[3] R. Winkel, Sequences of symmetric polynomials and combinatorial properties of tableaux. Adv. Math. 134(1998):46-89.

和 Kreimer[1][2]于 2000 年将 Rota-Baxter 代数引入到量子域重整规化的研究,它是量子域重整规化理论从代数角度研究的奠基性工作. 于是 20 世纪,Rota-Baxter 代数在数学物理等方面得到了巨大的发展.

Rota-Baxter 代数与数学、数学物理有着十分紧密的联系,而且 Rota-Baxter 代数的研究与其他有着丰富成果的数学领域相比,如微分代数,还尚处在研究的初级阶段,所以有着十分广泛的研究与发展前景[3][4][5].

定义 1 设 $(A,+,.)$ 是一个环,k 是一个域. 如果 $(A,+)$ 是一个 k-模,且对于任意 $r\in k$ 和 $a,b\in A$,均有 $(ra)b=a(rb)=r(ab)$,则称 A 是一个 k-代数. 若 k-代数间的映射 $f:A\rightarrow B$ 是环同态,且是 k-模同态,则称 f 是一个 k-代数同态.

定义 2 设 R 是一个 k-代数,如果 R 中的一个线性算子 $P:R\rightarrow R$ 满足 Rota-Baxter 方程

---

[1] A. Connes and D. Kreimer. Renormalization in quantum field theory and the Riemann-Hilbert problem. Ⅱ. The Hopf algebra structure of graphs and main theorem. Comm. Math. Phys. ,210(2000),no.1:249-273.

[2] A. Connes and D. Kreimer,Renormalization in quantum field theory and the Riemann-Hilbert problem. Ⅱ. The β-function, diffeomorphisms and the renormalization group. , Comm. Math. Phys. ,216(2001),no.1:215-241.

[3] L. Guo,An Introduction to Rota-Baxter Algebra. International Press(US) and Higher Education Press(China),2012.

[4] L. Guo and W. Keigher. On differential Rota-Baxter algebra. J. Pure Appl. Algebra, 212(2008):552-540.

[5] 文中的环 R,是有单位元 1_R 的交换环. 我们用 $\mathcal{N}$ 表示自然数集构成的加法幺半群,$\mathcal{N}_+$ 表示正整数构成的加法半群,$\mathcal{R}$ 表示实数域. 文中有关的概念和记号均参见 An Introduction to Rota-Baxter Algebra,L. Guo.

$$P(x)P(y) = P(xP(y)) + P(P(x)y) + \lambda P(xy), \forall x,y \in R \tag{1}$$

则称 P 是 R 上的一个权重为 λ 的 Rota-Baxter 算子(简称 RBO),其中 $\lambda \in k$.

定义 3 设 R 是一个 k-代数,P 是 R 上的一个权重为 λ 的 Rota-Baxter 算子,则称 (R,P) 为权重为 λ 的 Rota-Baxter 代数,简称为权重为 λ 的 RB 代数.

显然,0 映射 $0:R \to R$ 是任意环 R 上的 Rota-Baxter 算子.因此每一个 k-代数都可以看成是一个 Rota-Baxter k-代数.单位映射 I_p 显然是权重为 -1 的 Rota-Baxter 算子.

k-代数的基本概念可以类似地在 Rota-Baxter 代数中有相应的基本概念.特别地,设 $(R,P),(S,Q)$ 是两个权重为 λ 的 Rota-Baxter 代数,如果 k 代数同态 $f:(R,P) \to (S,Q)$ 满足 $f(P(x)) = Q(f(x))$,则称 f 是一个 Rota-Baxter 代数同态. Rota-Baxter 子代数(或 Rota-Baxter 理想)是 R 的一个子代数(或理想)I 满足 $P(I) \subseteq I$.如果 I 是 Rota-Baxter 代数 (R,P) 的 Rota-Baxter 理想,则 Rota-Baxter 商代数是商代数 R/I.如果 $f:(R,P) \to (S,Q)$ 是 Rota-Baxter 代数同态,则 $\mathrm{Ker} f$ 是 k-代数 R 的理想,而 $\mathrm{Im} f \subseteq Q$ 是 (S,Q) 的一个 Rota-Baxter 子 k-代数.

下面我们给出几个 Rota-Baxter 代数的例子.

例 1 (数乘)设 R 是一个 k-代数,给定 $\lambda \in k$. 定义算子 P 为

$$P:R \to R, P(x) = -\lambda x, \forall x \in R$$

则 P 为 R 上权重为 λ 的 Rota-Baxter 算子.特别地,恒等映射是权重为 -1 的 Rota-Baxter 算子

例2 （积分）设 R 是 $\mathscr{R}$ 上全体连续函数构成的 $\mathscr{R}$-代数. 定义算子 P 为

$$P:R\to R, P(f)(x) = \int_0^x f(t)\,\mathrm{d}t, \forall x \in \mathscr{R}$$

则 (R,P) 为权重为 0 的 Rota-Baxter 代数.

设 $F(x) = \int_0^x f(t)\,\mathrm{d}t, G(x) = \int_0^x g(t)\,\mathrm{d}t, f,g \in R$ 由分部积分法公式

$$\int_0^x F'(t)G(t)\,\mathrm{d}t = F(t)G(t)\big|_0^x - \int_0^x F(t)G'(t)\,\mathrm{d}t$$

即

$$\int_0^x f(t)G(t)\,\mathrm{d}t = F(x)G(x) - \int_0^x F(t)g(t)\,\mathrm{d}t$$

有

$$\int_0^x f(t)\,\mathrm{d}t \int_0^x g(t)\,\mathrm{d}t = \int_0^x f(t)\left(\int_0^t g(s)\,\mathrm{d}s\right)\mathrm{d}t + \int_0^x \left(\int_0^t f(s)\,\mathrm{d}s\right)g(t)\,\mathrm{d}t$$

即

$$P(f)(x)P(g)(x) = P(fP(g))(x) + P(P(f)g)(x)$$
$$P(f)P(g) = P(fP(g)) + P(P(f)g), \forall f,g \in R$$

例3 （均幂代数）设 $R = \bigoplus_{n\geq 0} Ce_n$，它的乘法为 $e_m e_n = \binom{m+n}{m}e_{m+n}, m,n \geq 0$. 定义 P 为

$$P:R\to R, P(e_n) = e_{n+1}, n\geq 0$$

则 (R,P) 为权重为 0 的 Rota-Baxter 代数.

例4 （Hurwitz 序列[①]）设 R 为

---

① W. Keigher, On the ring of Hurwitz series. Comm. Algebra, 25(1997):1845-1859.

$$HR:=\{(a_n)\mid a_n\in R, n\in \mathcal{N}\}$$

Hurwitz 序列环的加法为分量相加，即 $(a_n)+(b_n)=(a_n+b_n)$. 它的乘法类似于幂级数的乘法，即 $(a_n)(b_n)=(c_n)$，其中 $c_n=\sum_{k=0}^{n}\binom{n}{k}a_k b_{n-k}$. 定义算子 P 为

$$P:HR\to HR, P((a_n))=(a_{n-1}), \forall (a_n)\in HR, a_{-1}=0$$

则 (HR,P) 为权重为 0 的 Rota-Baxter 代数.

例5 （部分和[①]）设 A 是一个 k-代数，R 是取值在 A 中序列 $(a_n)_{n\geq 1}$ 组成的集合. R 的加法和乘法是普通向量的加法和标量积，则 R 构成一个 k-代数. 定义 P 为

$$P:R\to R, P(a_1,a_2,\cdots,a_n,\cdots)$$
$$=\lambda(0,a_1,a_1+a_2,a_1+a_2+a_3,\cdots), a_n\in A$$

则 P 为 R 上权重为 λ 的 Rota-Baxter 算子.

例6 定义 P 为

$$P:R\to R, P(a_1,a_2,\cdots,a_n,\cdots)=(a_1,a_1+a_2,\cdots,\sum_{1\leq i\leq n}a_i,\cdots)$$

则 P 为 R 上权重为 -1 的 Rota-Baxter 算子.

定义函数 $f:\mathcal{N}_+\to \mathcal{R}$，则 $P(f)$ 是部分和序列

$$P(f):\mathcal{N}_+\to \mathcal{R}, P(f)(n)=\sum_{1\leq i\leq n}f(i), n\geq 1$$

对于 $f,g\in R$，我们有

$$(P(f)P(g))(n)=P(f)(n)P(g)(n)$$
$$=(\sum_{1\leq i\leq n}f(i))(\sum_{1\leq j\leq n}g(j))$$
$$=(\sum_{1\leq i,j\leq n}f(i)g(j))$$

[①] G. Baxter, An analytic problem whose solution follows from a simple algebraic identity. Pacific J. Math. ,10(1960):731-742.

$$= \sum_{\mathrm{I}} f(i)g(j) = \sum_{\mathrm{II}} f(i)g(j) + \sum_{\mathrm{III}} f(i)g(j) - \sum_{\mathrm{IV}} f(i)g(j)$$

其中

$$\mathrm{I} := \{(i,j) \mid 1 \leq i \leq n, 1 \leq j \leq n\}$$
$$\mathrm{II} := \{(i,j) \mid 1 \leq j \leq i \leq n\}$$
$$\mathrm{III} := \{(i,j) \mid 1 \leq i \leq j \leq n\}$$
$$\mathrm{IV} := \{(i,j) \mid 1 \leq i = j \leq n\}$$

而

$$P(P(f)g)(n) = \sum_{1 \leq j \leq n} P(f)(j)g(j)$$
$$= \sum_{1 \leq j \leq n} \left(\sum_{1 \leq i \leq j} f(i) \right) g(i)$$
$$= \sum_{1 \leq i \leq j \leq n} f(i)g(j)$$
$$= \sum_{\mathrm{III}} f(i)g(j)$$

同理可证,$P(fP(g))(n) = \sum_{\mathrm{II}} f(i)g(j)$,$P(fg)(n) = \sum_{\mathrm{IV}} f(i)g(j)$,则 P 满足等式(1),此时 $\lambda = -1$.

下面我们由 Rota-Baxter 代数的定义讨论它的基本性质.

命题 1 (1) 设 (R,P) 是 Rota-Baxter 代数,则 $P(R)$ 是 R 的子代数.

(2) 如果 P 是权重为 λ 的 Rota-Baxter 算子,则 μP 是权重为 $\mu\lambda$ 的 Rota-Baxter 算子. 反之,若 P 是权重为 $\mu\lambda$ 的 Rota-Baxter 算子,且 μ 在 k 中可逆,则 $\mu^{-1}P$ 是权重为 λ 的 Rota-Baxter 算子.

证明 (1) 由等式(1)显然.

(2) 因为
$$P(x)P(y) = P(xP(y)) + P(P(x)y) + \lambda P(xy), \forall x,y \in R \quad (2)$$
则
$$(\mu P)(x)(\mu P)(y) = (\mu P)(x(\mu P)(y)) +$$
$$(\mu P)((\mu P)(x)y) +$$
$$\mu\lambda(\mu P)(xy), \forall x,y \in R \quad (3)$$
因此 μP 是权重为 $\mu\lambda$ 的 Rota-Baxter 算子.

反之,如果 μ 是可逆的,由式(3)我们得到式(2),从而得证.

命题 2 设 P 是权重为 $-\lambda$(或 λ)的 Rota-Baxter 算子,则 $\overline{P}$ 也是权重为 $-\lambda$(或 λ)的 Rota-Baxter 算子. 其中 $\overline{P} = \lambda I_P - P$(或 $\overline{P} = -\lambda I_P - P$),$I_P$ 是单位映射.

证明 如果 P 是权重为 $-\lambda$ 的 Rota-Baxter 算子,则 $\forall x,y \in R$,有
$$\overline{P}(x)\overline{P}(y) = (\lambda x - P(x))(\lambda y - P(y))$$
$$= \lambda^2 xy - \lambda x P(y) - \lambda P(x)y + P(x)P(y)$$
$$= -\lambda(xP(y) + P(x)y - \lambda xy) + P(xP(y) +$$
$$P(x)y - \lambda xy)$$
$$= -(\lambda I_P - P)(xP(y) + P(x)y + \lambda xy)$$
$$= -\overline{P}(xP(y) + P(x)y + \lambda xy)$$

另一方面
$$\overline{P}(x\overline{P}(y) + \overline{P}(x)y - \lambda xy)$$
$$= \overline{P}(x(\lambda y - P(y)) + (\lambda x - P(x))y + \lambda xy)$$
$$= -\overline{P}(xP(y) + P(x)y + \lambda xy)$$

$$= \overline{P}(x)\overline{P}(y)$$

所以,$\overline{P}$ 也是权重为 $-\lambda$ 的 Rota-Baxter 算子.

若 P 是权重为 λ 的 Rota-Baxter 算子,同理可证,$\overline{P} = -\lambda I_P - P$ 也是权重为 λ 的 Rota-Baxter 算子.

定理 3 设 P_1, P_2 是权重为 λ 的 Rota-Baxter 算子,且 P_1, P_2 满足

$$P_1(x)P_2(y) = P_1(xP_2(y)) + P_2(P_1(x)y)$$

则 $\widetilde{P} = P_1 + P_2$ 也是权重为 λ 的 Rota-Baxter 算子.

证明 如果 P_1, P_2 是权重为 λ 的 Rota-Baxter 算子. 则 $\forall x, y \in R$,有

$$\widetilde{P}(x)\widetilde{P}(y) = (P_1(x) + P_2(x))(P_1(y) + P_2(y))$$
$$= P_1(x)P_1(y) + P_2(x)P_2(y) + P_1(x)P_2(y) + P_2(x)P_1(y)$$
$$= P_1(xP_1(y) + P_1(x)y + \lambda xy) + P_2(xP_2(y) + P_2(x)y + \lambda xy) + P_1(xP_2(y)) + P_2(P_1(x)y) + P_2(xP_1(y)) + P_1(P_2(x)y)$$
$$= \widetilde{P}(xP_1(y)) + \widetilde{P}(P_1(x)y) + \widetilde{P}(xP_2(y)) + \widetilde{P}(P_2(x)y) + \widetilde{P}(\lambda xy)$$
$$= \widetilde{P}(xP_1(y) + xP_2(y) + P_1(x)y + P_2(x)y + \lambda xy)$$

另一方面

$$\widetilde{P}(x\widetilde{P}(y) + \widetilde{P}(x)y + \lambda xy)$$
$$= \widetilde{P}(x(P_1(y) + P_2(y)) + (P_1(x) + P_2(x))y + \lambda xy)$$
$$= \widetilde{P}(xP_1(y) + xP_2(y) + P_1(x)y + P_2(x)y + \lambda ry)$$

所以,$\widetilde{P}$ 也是权重为 λ 的 Rota-Baxter 算子.

推论 4 设 P_1,P_2 是权重分别为 λ 和 $-\lambda$ 的 Rota-Baxter 算子,且满足如上条件,则 $\widetilde{P}=P_1-P_2$ 也是权重为 λ 的 Rota-Baxter 算子.

定理 5 设 R 是一个 k-代数,且 $R=R_1\oplus R_2$ 是 R 上 k-横直和分解,R_1 是单位的 k-子代数. 如果 R 中的一个线性算子 P 满足
$$P:R\to R_2\subseteq R, P(r_1+r_2)=r_2, \forall r_1\in R_1, r_2\in R_2$$
则满射 P 是权重为 -1 的幂等的 Rota-Baxter 算子.

证明 设 R 有 k-模直和分解 $R=R_1\oplus R_2$. 如果 R 的算子 P 将 R 满射到 R_2,则对于 $r=r_1+r_2, \forall r_1\in R_1, r_2\in R_2$,我们有
$$P^2(r)=P^2(r_1+r_2)=P(r_2)=r_2=P(r)$$
则 P 是幂等的.

设 $r=r_1+r_2, s=s_1+s_2$,其中 $r_1,s_1\in R_1, r_2,s_2\in R_2$. 因为 $r_1,s_1\in R_1, r_2,s_2\in R_2$,所以 $P(r_1s_1)=0, P(r_2s_2)=r_2s_2$. 因此
$$P(r)P(s)=r_2s_2$$
$$P(rP(s))=P((r_1+r_2)s_2)=P(r_1s_2+r_2s_2)=P(r_1s_2)+r_2s_2$$
$$P(P(r)s)=P(r_2(s_1+s_2))=P(r_2s_1+r_2s_2)=P(r_2s_1)+r_2s_2$$
$$P(rs)=P((r_1+r_2)(s_1+s_2))=P(r_1s_1+r_1s_2+r_2s_1+r_2s_2)$$
$$=P(r_1s_2)+P(r_2s_1)+r_2s_2$$

我们得到
$$P(r)P(s)=P(rP(s))+P(P(r)s)-P(rs)$$
因此,P 是权重为 -1 的幂等的 Rota-Baxter 算子.

周教授还得到如下的定理：

定理6 设 R 是一个 k-代数，R 上的线性算子 P 是权重为 -1 幂等的 Rota-Baxter 算子当且仅当存在 R 非酉的 k-子代数 R_1, R_2 的 k-模直和分解 $R = R_1 \oplus R_2$，使得
$$P: R \to R_1 \subseteq R \tag{4}$$
是 R 到 R_1 上的满射，满足 $P(a_1 + a_2) = a_1$，$\forall a_1 \in R_1, a_2 \in R_2$。

证明 如果 R 到非酉的 k-子代数 R_1, R_2 有 k-模直和分解 $R = R_1 \oplus R_2$，则对于 $a = a_1 + a_2$，$\forall a_1 \in R_1, a_2 \in R_2$，我们有
$$P^2(a) = P^2(a_1 + a_2) = P(P(a_1 + a_2)) = P(a_1) = a_1 = P(a)$$
则 P 是幂等的。

设 $a = a_1 + a_2, b = b_1 + b_2$，其中 $a_1, b_1 \in R_1, a_2, b_2 \in R_2$。因为 $a_1 b_1 \in R_1, a_2 b_2 \in R_2$，所以 $P(a_1 b_1) = a_1 b_1, P(a_2 b_2) = 0$。因此

$$P(a)P(b) = a_1 b_1$$
$$P(aP(b)) = P((a_1 + a_2) b_1)$$
$$= P(a_1 b_1 + a_2 b_1) = a_1 b_1 + P(a_2 b_1)$$
$$P(P(a)b) = P(a_1(b_1 + b_2)) = P(a_1 b_1 + a_1 b_2)$$
$$= a_1 b_1 + P(a_1 b_2)$$
$$P(ab) = P((a_1 + a_2)(b_1 + b_2))$$
$$= P(a_1 b_1 + a_1 b_2 + a_2 b_1 + a_2 b_2)$$
$$= a_1 b_1 + P(a_1 b_2) + P(a_2 b_1)$$

我们得到 $P(a)P(b) = P(aP(b)) + P(P(a)b) - P(ab)$。

因此，P 是权重为 -1 的幂等的 Rota-Baxter 算子。

反之,设 P 是权重为 -1 幂等的 Rota-Baxter 算子. 令 $R_1 = P(R)$ 且 $R_2 = (I_P - P)(R)$,得 R_1, R_2 是 R 非酉的 $k-$子代数,且 $a = P(a) + (I_P - P)(a)$, $\forall a \in R$.

因此 $R = R_1 + R_2$. 如果 $a \in R_1 \cap R_2$,则 $a = P(a_1) = (I_P - P)(a_2), a_1, a_2 \in R$.

所以 $a = P(a_1) = P^2(a_1) = P(L_P - P)(a_2) = (P - P^2)(a_2) = 0$.

从而 $R = R_1 \oplus R_2$. 又因为 $a = P(a) + (I_P - P)(a)$ 是 $a = a_1 + a_2, a_1 \in R_1, a_2 \in R_2$ 的分解,所以 P 是 R 到 R_1 上的满射.

定理 7 设 (R, P) 是一个 Rota-Baxter 代数,且 P 是权重为 λ 的幂等的 Rota-Baxter 算子,则

$$(1 + \lambda)P(P(x)y) = 0 \qquad (5)$$

$$(1 + \lambda)P(xP(y)) = 0 \qquad (6)$$

$$(1 + \lambda)(P(x)P(y) - \lambda P(xy)) = 0, \forall x, y \in R \qquad (7)$$

因此,如果 k 是一个域,则 (R, P) 是一个权重为 -1 的 Rota-Baxter 代数.

证明 由于 (R, P) 为权重为 λ 的 Rota-Baxter 代数,对于 $\forall x, y \in R$,由定义有

$$P(x)P(y) = P(xP(y) + P(x)y + \lambda xy) \qquad (8)$$

由于 P 是幂等的,因此

$P(x)P(y)$
$= P^2(x)P(y)$
$= P(P(x)P(y) + P^2(x)y + \lambda P(x)y)$
$= P(P(x)P(y)) + P(P^2(x)y) + \lambda P(P(x)y)$
$= P^2(xP(y) + P(x)y + \lambda xy) + P(P(x)y) + \lambda P(P(x)y)$

$$= P(xP(y) + P(x)y + \lambda xy) + (1+\lambda)P(P(x)y)$$

因此,式(5)成立. 同理可证式(6). 由式(5)(6)(8), 得

$$(1+\lambda)(P(x)P(y) - \lambda P(xy))$$
$$= (1+\lambda)(P(xP(y) + P(x)y + \lambda xy) - \lambda P(xy))$$
$$= (1+\lambda)(P(xP(y)) + P(P(x)y) + \lambda P(xy) - \lambda P(xy))$$
$$= (1+\lambda)(P(xP(y)) + P(P(x)y))$$
$$= (1+\lambda)P(xP(y)) + (1+\lambda)P(P(x)y) = 0$$

下面我们给出 Rota-Baxter 代数在组合学中的应用.

例 7(Maximum) 设函数 $m: R \to R, x \to x^+ := \max\{0, x\}, x \in R$.

这里 R-代数为 $R := \{f: R \to Q \mid |\text{supp}(f)| < \infty\}$,其中 $\text{supp}(f) := \{x \in R \mid f(x) \neq 0\}$,且定义它们的积为

$$(fg)(x) := \sum_{y, z \in R, y+z=x} f(y)g(z), x \in R$$

以及定义算子为 $P: R \to R, P(f)(x) := \sum_{y \in R, y+z=x} f(y) = \sum_{\max\{0, y\}=x} f(y), x \in R$.

规定在空集上其和为 0,即

$$p(f)(x) = \begin{cases} 0, x < 0 \\ \sum_{y \leq 0} f(y), x = 0 \\ f(x), \text{当 } x > 0 \end{cases} \quad (9)$$

则有 $\text{supp}(P(f)) \subseteq R_{\geq 0}$. 由式(6),如果 $\text{supp}(f) \subseteq R_{\geq 0}$,则 $P(f) = f$. 从而

$$P(R) = R_1 := \{f \in R \mid \text{supp}(f) \subseteq R_{\geq 0}\}$$

对于 $f, g \in R_1$,由于 $(fg)(x) := \sum_{y+z=x} f(y)g(z)$,当 $x < 0$ 时,由 $y + z = x$ 得 $y < 0$ 或 $z < 0$,则 $f(y)g(z) = 0$. 因此 $\mathrm{supp}(fg) \subseteq R_{\geq 0}$,从而 R_1 是 R 的子代数,定义 $P(R) = R_2 := \{f \in R \mid \mathrm{supp}(f) \subseteq R_{\leq 0}, \sum_{x \in R} f(x) = 0\}$,则 R_2 是 R 的子代数. 对于 $f \in R$,有 $f = f_1 + f_2$,其中 $f_1(x) := P(f)(x)$ 且

$$f_2(x) = \begin{cases} 0, x < 0 \\ -\sum_{y \leq 0} f(y), x = 0 \\ f(x), x > 0 \end{cases} \quad (10)$$

这样 $R = R_1 + R_2$,显然 $R_1 \cap R_2 = 0$. 因此 $R = R_1 \oplus R_2$,由定理 3,(R, P) 为权值为 -1 的 Rota-Baxter 代数.

本书的第四章涉及可积系统.

可积系统是理论物理学最活跃的研究领域之一,在几十年的过程中发展了一系列成熟的理论方法,极大地推动了物理学的发展. 量子群与量子对称性是可积系统研究的最新发展,在广泛的物理问题中有重要的应用. 实际上,量子对称性已经超越可积系统的原来框架,也已成为解决不可积问题的重要手段. 中国科学院高能物理研究所的常哲研究员,1997 年撰文评述了量子群的引入背景、量子群理论,特别是量子对称性研究的最新结果[①].

自然界中到处都有对称性. 对很多物体来说,左右对称性有时是明显的. 也许人类的祖先很早就注意到了我们生活中的对称性. 但是,利用对称性来确定物质运动行

---

① 摘自《中国科学基金》,1997 年,第 4 期.

为的历史并不是很长,只是在现代物理学中对称性才确立了其无可替代的地位.在近代物理学中,对称性与可观测量或不可观测量联系在一起.说一系统具有时空平移不变性,等同于该系统能量及动量守恒,而旋转不变的系统一定是角动量守恒的.由于物质运动的复杂性,很难了解物质运动的详细情况,在有些情况下甚至是不可能的,如基本粒子内部的夸克运动等.这时,物理学家的唯一武器也许就是对称性了.漫步于今天的物理学无处不看到对称性的影子,可以不过分地说,现代物理学是伴随着对称性的研究而发展的.点群的分类与表示是固体物理学的基础;共形不变性是理解临界现象与超弦理论的关键;没有么正么模群就根本谈不上高能物理的所谓标准模型.但是,无论所研究的系统多么复杂多变,从数学结构上讲这些对称性研究的工具都是群论.

近年来,一种新的对称性——量子对称性及其数学结构量子群的研究波及物理与数学界的方方面面.而量子群实际上并非群甚至连半群都谈不上,只是由于历史的原因才被冠之以量子群这一屡遭误解的名字.量子群的引入与可积系统的研究紧密联系在一起.

1967 年,杨振宁先生[1]仔细地研究了一类具有 δ 势的一维量子场论问题.利用 Bethe Ansatz 方法,杨振宁先生证明,如果算子 $R(u)$ 满足某种关系式就可以得到这类问题的严格解.Baxter 教授在有关八顶角模型的论文中[2]得到关于顶角的玻尔兹曼权重因子 $w(\beta_i,\alpha_i,\beta_i,\alpha_i|u)$ 的一

---

[1] Yang C N. Phy. Rev. Lett. ,1967,19:1312.

[2] Baxter R J. Ann. Phys. (N.Y.) ,1972,70:193.

组关系式. 如果把 $R(u)$ 与 $w(\beta,\alpha,\beta',\alpha'|u)$ 理解为张量乘积空间 $V^{\otimes 2}$ 的算子 $\mathscr{R}(u)$,则在 $V^{\otimes 3}$ 中可以把它们统一写成

$$\mathscr{R}_{12}(u)\mathscr{R}_{13}(u+v)\mathscr{R}_{23}(v) = \mathscr{R}_{23}(v)\mathscr{R}_{13}(u+v)\mathscr{R}_{12}(u)$$

这就是通常形式的 Yang-Baxter 方程. Yang-Baxter 方程一般表示为系统的可积条件. 一旦得到一组 Yang-Baxter 方程解,问题也就随之得到解决. 从而,Yang-Baxter 方程成为可积系统的主导方程,求解 Yang-Baxter 方程也就成为研究可积问题的重要课题. 以致有人评论说,20 世纪 70 年代 Yang-Mills 理论改变了整个高能物理理论,Yang-Baxter 方程则构成 20 世纪 90 年代理论物理学最重要的研究课题. Yang-Mills 理论与 Yang-Baxter 方程是杨振宁先生对物理学的两个划时代的重要贡献.

  Yang-Baxter 方程是一组非线性算子方程,求解非常困难,杨振宁先生得到了它的第一解,之后经过长期努力又求得一系列个别解. 20 世纪 80 年代以后,Drinfeld 与 Jimbo 等人[1]开始探讨 Yang-Baxter 方程的系统解方法. 人们注意到,Yang-Baxter 方程的一类重要解包含一个额外参数 γ(量子化参数),如果在量子化参数的零点附近对这类解作展开,就得到 $\mathscr{R}(u,\gamma)$ 的经典极限 $r(u)$ 满足的一个较简单的方程——经典 Yang-Baxter 方程. 利用李代数的表示论可以得到一类经典 Yang-Baxter 方程解的完整分类. 正是这一成功促使人们构造一种新的数学结构,它的表示可以给出 Yang-Baxter 方程的系统解. 这也就是

---

[1] Drinfeld V G. in Proceedings of the ICM, Berkeley 1986, ed. Gleason A. M. (American Mathematical Society, Providence, RI, 1987) p798; Jimbo M. Lett. Math. Phys., 1985, 10: 63.

"量子群"命名的历史缘由.一旦引进了量子群理论,Yang-Baxter 方程的研究也就突破了可积系统的范畴,成为物理与数学界广泛关注的问题.

域 C 上的一个代数是包含两个线性运算:乘法 m 与单位 η 的线性空间 A,$m:A\otimes A\to A$,$\eta:C\to A$,并满足结合律,存在单位元.当涉及表示张量积时还要引进另外一个线性运算:余乘积 $\Delta:A\to A\otimes A$.余乘积为满足余结合律的代数同胚.如果再引进余单位运算 $\varepsilon:A\to C$,满足 $(\mathrm{id}\otimes\varepsilon)\Delta=(\varepsilon\otimes\mathrm{id})\Delta=\mathrm{id}$,则我们构成一个双代数.与每个群元素都有逆这样一个事实相联系,有些双代数有一个被称为 antipode 的运算 $S:A\to A$. Antipode 满足 $m(S\otimes\mathrm{id})\Delta=m(\mathrm{id}\otimes S)\Delta=\eta\circ\varepsilon$.这样一类双代数被称为 Hopf 代数.一个李代数 g 的包络 $U(g)$,如果定义乘法为普通乘法,$\Delta(x)=x\otimes 1+1\otimes x$,$\eta(\alpha)=\alpha 1$,$\varepsilon(1)=1$,$\varepsilon(x)=0(x\neq 1)$,$S(x)=-x$,则构成一个 Hopf 代数.

李双代数是一个具有满足 1-cocycle 条件的余反对称运算 ψ 的李代数.李双代数 $(A_{(0)},m_{(0)},\Delta_{(0)},\eta_{(0)},\varepsilon_{(0)},\psi)$ 的量子化是环 $C[[\gamma]]$ 上的非余交换代数 $(A,m,\Delta,\eta,\varepsilon)$.一个通常的李单代数或非扭曲仿射 Kac-Moody 代数 g 的 Borel 包络子代数 $U^+(g)$ 与 $U^-(g)$ 的量子化对应被记作 $U_q^+(g)$ 与 $U_q^-(g)$.对一个 Hopf 代数 $(A,m,\Delta,\eta,\varepsilon,S)$ 可以用递推法定义高阶余乘法 $\Delta^{(n-1)}:A\to A^{\otimes n}[\Delta^{(1)}=\Delta,\Delta^{(n)}=(\Delta\otimes\mathrm{id})\circ\Delta^{(n-1)}]$.一般情况下,对两个 Hopf 代数 A 与 B 及满足下列条件的非退化双线性形式 $(\langle\cdots,\cdots\rangle:A\times B\to C)$,有

$$\begin{cases} \langle a^i, b_j b_k \rangle = \langle \Delta_A(a^i), b_j \otimes b_k \rangle \\ \langle a^i a^j, b_k \rangle = \langle a^j \otimes a^i, \Delta_B(b_k) \rangle \\ \langle 1^A, b_j \rangle = \varepsilon_B(b_i), \langle a^i, 1_B \rangle = \varepsilon_A(a^i) \\ \langle S_A(a^i), S_B(b_j) \rangle = \langle a^i, b_j \rangle \end{cases} \quad (1)$$

可以证明如下定理：存在唯一的$(A, B, \langle \cdots, \cdots \rangle)$的量子偶$D$，满足：

A与B都是D的Hopf子代数；

如果$\{a^\alpha\}$与$\{b_\beta\}$分别是A与B的基，那么乘积$\{a^\alpha b_\beta\}$则构成D的一组基；

对于$a^i \in A, b_j \in B$，和$\sum a^{i(1)} \otimes a^{i(2)} \otimes a^{i(3)} \equiv \Delta_A^{(2)}(a^i)$，$\sum b_{j(1)} \otimes b_{j(2)} \otimes b_{j(3)} \equiv \Delta_B^{(2)}(b_j)$，有$b_j a^i = \sum \langle a^{i(1)}, S(b_{j(1)}) \rangle \langle a^{i(3)}, b_{j(3)} \rangle a^{i(2)} b_{j(2)} \in D.$

$(U_q^+(g), U_q^-(g), \langle \cdots, \cdots \rangle)$的量子偶$D$加上约束$q_i^{2H_i}[\in U_q^+(g)] = q_i^{2H_i}[\in U_q^-(g)]$就得到了量子群$U_q(g)$[3]

$$\begin{cases} [H_i, H_j] = 0, [H_i, X_j^\pm] = \pm a_{ij} X_i^\pm \\ [X_i^+, X_j^-] = \delta_{ij}[H_i]_{qi} \\ [X_i^\pm, X_j^\pm] = 0 (a_{ij} = 0) \\ \sum_{m=0}^{1-a_{ij}} (-1)^m \begin{bmatrix} 1-a_{ij} \\ m \end{bmatrix}_{qi} (X_i^\pm)^{1-a_{ij}-m} X_j^\pm (X_i^\pm)^m = 0 \quad (i \neq j) \\ \Delta(H_i) = H_i \otimes 1 + 1 \otimes H_i \\ \Delta(X_i^\pm) = X_i^\pm \otimes q_i^{H_i} + q_i^{-H_i} \otimes X_i^\pm \\ \varepsilon(H_i) = 0 = \varepsilon(X_i^\pm), S(H_i) = -H_i \\ S(X_i^\pm) = -q^{-\Sigma_i H_i} X_i^\pm q^{\Sigma_i H_i} \end{cases} \quad (2)$$

这里我们使用了记号 $q_i = q^{d_i}$, $\begin{bmatrix} m \\ n \end{bmatrix}_q = \frac{[m]_q!}{[n]_q! [n-m]_q!}$,

$[m]_q = \frac{q^m - q^{-m}}{q - q^{-1}}$.

除了晶体的点群对称性等静止对称性以外,我们了解的大多数物理对称性是运动系统的对称性. 李政道先生在谈到运动系统的对称性时曾形象地拿一支铅笔在一张白纸上左右晃动而阐述左右的大致对称性. 实际的物理对称性很少有绝对的对称性. 新的研究方法可以发现新的对称性. 某种对称性破缺的系统却可能存在另外一种对称性,这也正是对称性研究的魅力所在. 量子群的引进给物理学家研究新的物理系统对称性提供了有力的工具. 一般称系统的量子群对称性为量子对称性. 量子对称性的研究使得量子群从单纯 Yang-Baxter 系统解构造的工具而转变为广泛物理领域关注的焦点.

一类经典系统——纯自旋系统(pure-spin systems)可以由相对刚体标架坐标的角动量分量 J_i 唯一地确定. 对这类系统,利用广义泊松括号及导数规则可以得到标准哈密顿形式的运动方程[①].

变形对称陀螺的哈密顿量为

$$H_q = \frac{I - I_3}{2II_3} J'^2_3 + \frac{1}{2I}\left(J'^2_1 + J'^2_2 + \frac{(\sinh \gamma J'_3)^2}{\gamma \sinh \gamma}\right), q = e^\gamma \quad (3)$$

它的相空间与辛形式由下式给出

$$M\delta : J'^2_1 + J'^2_2 + \frac{(\sinh \gamma J'_3)^2}{\gamma \sinh \gamma} = J_q^2, J_q = \frac{\sinh \gamma J_0}{\sqrt{\gamma \sinh \gamma}}$$

---

① Chang Z. Phys. Rev., 1992, A45:4303; idid. 1992, A46:1400.

$$\Omega_q = \frac{1}{J_q^2}\left(J'_1 dJ'_2 \wedge dJ'_3 + J'_2 dJ'_3 \wedge dJ'_1 + \frac{\tanh \gamma J'_3}{\gamma} dJ'_1 \wedge dJ'_2\right)$$

标准过程给出基本广义泊松括号

$$[J'_+, J'_-]_{GPB} = i\frac{\sinh 2\gamma J'_3}{\sinh \gamma}, [J'_3, J'_\pm]_{GPB} = \pm i J'_\pm \quad (4)$$

对上面的代数关系采用预量子化方法可以引进 Hopf 运算,这也就是所谓的 $SU_{q,h \to 0}(2)$ 量子群[①],它给出人们熟悉的对称陀螺运动方程.

利用几何量子化方法,我们可以把上面的讨论推广到量子对称陀螺系统. 形变量子对称陀螺的哈密顿量与其本征值为

$$\begin{cases} \hat{H}_q = \frac{\hbar^2(I-I_3)}{2II_3}\hat{J}'^2_3 + \frac{\hbar^2}{2I}\left(\frac{\gamma}{\sinh\gamma}\hat{J}'_+\hat{J}'_- + [\hat{J}'_3]_q[\hat{J}'_3+1]_q\right) \\ E^q_{JK} = \frac{\hbar^2}{2I}[J]_q[J+1]_q + \frac{I-I_3}{2II_3}\hbar^2 K^2 \end{cases}$$

(5)

量子形变陀螺的能量本征值有非常明确的物理含义,形变参数 q 对应系统的非线性修正. 这对陀螺分子的精细光谱结构给出了一种可能的解释.

六顶角模型的 Yang-Baxter 方程谱参数解 $\mathscr{R}(u)$,使得转移矩阵 $v^{(n)}(u)$ 可作为守恒量的产生泛函[②]. 而 $v^{(n)}(u)$ 的作用空间 $V^{\otimes m}$ 可认定为一维量子系统的希尔伯特空间. $v^{(n)}(u)$ 与 $v^{(n)}(v)$ 的对易性导致无穷多量 $\mathscr{H}_1[\equiv$

---

① Chang Z. Chen W, Guo H Y. J. Phys., 1990, A23:4185.
② Saleur H, Zuber J-B. in String Theory and Quantum Gravity, eds Green M et al, World Scientific, Singapore, 1991.

$\frac{\partial^i}{\partial u^i}\log \nu^{(n)}(u)|_{u=0}$] 相互对易. 如果令 $\mathscr{H}_1$ 为该量子系统的哈密顿量,我们得到与 $\mathscr{H}_1$ 对易的无穷多守恒量. 这种方式得到的哈密顿量形式为

$$\begin{cases} \mathscr{H}_1^2 = -\frac{1}{\sin \eta}\sum_{i=1}^n e_i \\ e_i = 1_{(1)}\otimes 1_{(2)}\otimes\cdots\otimes 1_{(i-1)}\otimes e_{(i,i+1)}\otimes 1_{(i+2)}\otimes\cdots\otimes 1_{(n)} \\ e = -\frac{1}{2}(\sigma^1\otimes\sigma^1 + \sigma^2\otimes\sigma^2 + \cos\eta\sigma^3\otimes\sigma^3) + \\ \quad \frac{1}{2}\cos\eta\cdot 1 - \frac{i}{2}\sin\eta(\sigma^3\otimes 1 - 1\otimes\sigma^3) \end{cases}$$

(6)

这里 η 是自由参数 ($q = e^{i\eta}$), $\boldsymbol{\sigma}^i$ 为泡利矩阵, e_i 是 $1 - P_{i(i+1)}$ 的量子对应, 满足所谓的 Temper-ley-Lieb 代数. 定义 $S^\pm = \Delta^{(n)}(s^\pm), S^3 = \Delta^{(n)}(s^3)$, 不难证明哈密顿量 $\mathscr{H}_1$ 是量子群 $SU_q(2)$ 不变的, 即 $[\mathscr{H}_1, SU_q(2)] = 0$. 从而顶角模型除了保证自旋守恒的 $U(1)$ 对称性以外还具有 $SU_q(2)$ 量子群对称性. 按同样的方法可以证明 S.O.S 模型的 diagonal-to-diagonal 转移矩阵给出与顶角模型同样的哈密顿量. 从而我们可以得到这样的结论, 可积格点模型是量子群不变的. 量子群生成元作用在格点的对角线上, 其可以解释为量子群自旋 $-1/2$ 表示的直积. 量子对称性方法把可积格点模型的解映射为一组耦合代数方程[①]. 这就使得求解格点模型变得简单易行.

---

① Levy D, Phys. Rev. Lett., 1990, 64:499; ibid., 1991, 67:1971; Hou B Y, Hou B Y, Ma Z Q. J. Phys., 1991 A24:2847.

共形场论最小模型的库仑气体形式由无穷远处存在背景荷 $2\alpha_0$ 的无质量标量场 ϕ 给出. 利用屏蔽流 $J(z)$, 定义[1]屏蔽算子 $X_\pm^-$, 有

$$X_\pm^- V_\alpha(z) = \frac{1}{1-q_\pm^{-1}} \int_C \mathrm{d}t J_\pm V_\alpha(z) \qquad (7)$$

这里 $q_\pm = \mathrm{e}^{2\pi \mathrm{i} \alpha_\pm^2}$, 适当选择的积分路径使得屏蔽算子为一 intertwiner. 再引进算子 $k_\pm$

$$k_\pm V_\alpha(z) = \exp(-2\mathrm{i}\pi\alpha_\pm \oint_c \partial\phi) V_\alpha(z)$$

可以自然得到最小模型对应的量子群 Borel 子代数

$$\begin{cases} [X_+^-, X_-^-] = 0, [k_+, k_-] = 0 \\ k_\pm X_\pm^- k_\pm^{-1} = q_\pm^{-1} X_\pm^-, k_\pm X_\mp^- k_\pm^{-1} = -X_\mp^- \end{cases} \qquad (8)$$

我们利用屏蔽算子在 Virasoro 代数下的变换性质定义 $X_\pm^-$ 的对偶 $X_\pm^+$. $X_\pm^-$ 与 $k_\pm$ 给出最小模型量子群结构的另外一个 Borel 子代数

$$\begin{cases} [X_+^+, X_-^+] = 0, k_\pm X_\pm^+ k_\pm^{-1} = q_\pm X_\pm^+, k_\pm X_\mp^+ k_\pm^{-1} = -X_\mp^+ \\ X_\pm^+ X_\pm^- - q_\pm X_\pm^- X_\pm^+ = \dfrac{1-k_\pm^{-2}}{1-q_\pm^{-1}}, X_\pm^+ X_\mp^- - q_\pm X_\mp^- X_\pm^+ = 0 \end{cases}$$

(9)

定义算子 $X_\pm^-$ 在普通乘积空间 $V_\alpha(z_1) V_\beta(z_2)$ 的作用为 Hopf 代数的余乘运算. $k_\pm$ 的余乘运算由屏蔽流 $J_\pm$ 与顶角算子 $V_\alpha(z)$ 的算子性质给出. 而由 Virasoro 算子可以得到 $X_\pm^+$ 的余乘运算

---

[1] Gómez C, Sierra G. Nucl. Phys., 1991, B352: 791; Ramírez C, Ruegg H, Ruiz-Altaba M. ibid., 1991, B364: 195.

$$\begin{cases} \Delta(X_\pm^-) = X_\pm^- \otimes 1 + k_\pm^{-1} \otimes X_\pm^- \\ \Delta(k_\pm) = k_\pm \otimes k_\pm, \Delta(X_\pm^+) = X_\pm^+ \otimes 1 + k_\pm^{-1} \otimes X_\pm^+ \end{cases} (10)$$

余乘以外的其他 Hopf 运算形式为

$$\begin{cases} \varepsilon(k_\pm) = 1, \varepsilon(X_\pm^-) = 0 = \varepsilon(X_\pm^+) \\ S(k_\pm) = k_\pm^{-1}, S(X_\pm^-) = -k_\pm X_\pm^-, S(X_\pm^+) = -k_\pm X_\pm^+ \end{cases}$$
(11)

Antipode 实质上是一个路径反运算,而余单位为一路径湮灭映射. $k_\pm, X_\pm^-$ 与 $X_\pm^+$ 生成的量子群代表了最小模型的量子对称性结构. 利用几乎完全相同的方法,我们可以研究 WZNW 模型的更一般的量子对称性结构.

考虑振转耦合后,一般形式的双原子分子振转谱由 Dunham 展开式[1]描述

$$E_{\text{vib-rot}}(v, J) = hc \sum_{ij} Y_{ij} (v + \frac{1}{2})^i (J(J+1))^j \quad (12)$$

一个成功的分子谱理论至少应该正确地导出 Dunham 展开式的几个主导项,给出分子波函数的形式,并有严格的跃迁选择定则. 事实上,同时满足这几个条件并非一件容易的事情[2].

具有量子对称性的振动谱由哈密顿量 $H_{q\text{-vib}}[= \frac{1}{2}(a_q^\dagger a_q + a_q a_q^\dagger) hc\nu_{\text{vib}}]$ 描述. 形变谐振子满足量子谐振子群 $H_q(4)$. 由量子谐振子群表示可以得到分子振动波函

---

[1] Herzberg G. Molecular Spectra and Molecular Structure. Van Nostrand, Princeton, 1955.

[2] Chang Z, Guo H Y, Yan H. Phys. Lett., 1991, A156:192; Chang Z, Yan H. ibid., 1991, A154:254; Bonatsos D et al. Chem. Phys. Lett., 1990, 175:300; Chang Z, Yan H. Phys. Rev., 1991, A44:7450.

数 $\widehat{\psi}_v(x) = N_v H_v(X) e^{-\frac{X^2}{2}}$. 该波函数给出振动谱的严格跃迁选择定则. 由量子谐振子群得到的振动谱

$$E_{q-\text{vib}}(v) = \frac{1}{2}([v+1+b\gamma]_q + [v+b\gamma]_q)hc\nu_{\text{vib}}$$

$$= hc\nu_{\text{vib}}\frac{\sinh(\gamma c)}{2\sinh(\frac{\gamma}{2})} + \cosh(\gamma c)(v+\frac{1}{2}) +$$

$$\frac{\gamma\sinh(\gamma c)}{2}(v+\frac{1}{2})^2 + \frac{\gamma^2\cosh(\gamma c)}{6}(v+$$

$$\frac{1}{2})^3 + \cdots, c \equiv b\gamma \qquad (13)$$

具量子对称性的转动哈密顿量为 $H_{q-\text{rot}} = \frac{h^2}{8\pi^2 I}(J'_- J'_+ + [J'_3]_q \cdot [J'_3 + 1]_q)$. 量子群 $SU_q(2)$ 的表示给出转动谱的波函数 $\widehat{\psi}_{JM}(x) = Y_{JM}(\theta, \phi)$. 此波函数给出严格的转动跃迁选择定则. 而转动谱由 Casimir 算子在该表示中的取值得到

$$E_{q-\text{rot}} = \frac{h^2}{8\pi^2 I}[J]_q[J+1]_q$$

$$= \frac{h^2}{8\pi^2 I}\left(\left(1 - \frac{1}{6}\gamma^2 + \frac{7}{360}\gamma^4\right)J(J+1) + \right.$$

$$\gamma^2\left(\frac{1}{3} - \frac{7}{90}\gamma^2\right)(J(J+1))^2 +$$

$$\left.\frac{2\gamma^4}{45}(J(J+1))^3 + \cdots\right) \qquad (14)$$

振转耦合谱由下面具量子对称性的哈密顿量描述

$$H_{q-\text{vib-rot}} = \frac{1}{2}(a^{\dagger}_{q(J)}a_{q(J)} + a_{q(J)}a^{\dagger}_{q(J)})hc\nu_{\text{vib}} +$$

$$\frac{h^2}{8\pi^2 I}(J_- J_+ + J_3(J_3 + 1)) \qquad (15)$$

量子化参数 q 不再是一个常数而是依赖于 J 的函数. 量子群直积 $H_{q(J)}(4) \otimes SU(2)$ 的表示给出振转耦合波函数 $\tilde{\psi}_{q(J)-\text{vib}-\text{rot}}(v,x) = N_v Y_{JM}(\theta,\phi) H_v(X) e^{-\frac{X^2}{2}}$. 严格的振转跃迁选择定则由该波函数给出. 量子对称性方法给出的振转耦合谱为

$$E = E_0 + \nu_{\text{vib}} \frac{1}{2\sinh(\gamma(J)/2)} \sinh(\gamma(J)(v + \frac{1}{2} + c(J))) + \frac{h^2}{8\pi^2 I} J(J+1) \qquad (16)$$

这样,我们就得到了双原子分子振转谱的量子对称性描述.

自量子群理论建立至今不过十年多的时间,物理学家与数学家们从不同的侧面对量子群理论进行了广泛的研究[①]. 量子群的表述形式也有多种. 对这样一个正在迅速发展的理论做详尽的评述几乎是不可能的. 这里,我们尽显简单地用也许不太严格的语言向范围比较广泛的读者介绍量子群的引入背景与研究量子对称性所必须的量子群理论,几乎没有讨论量子群的表示论,特别是量子化参数为单位根时的表示论,是件令人遗憾的事情. 量子群理论建立不久,物理学家就注意到量子对称性研究的潜在价值. 即使最初的热潮过去之后,依然可以看到各方面取得的不断进展. 正是量子对称性的研究使得量子群理论突破了可积系统的原来框架,而被赋予各种不同的物

---

① Chang Z. Physics Reports,1996,262:137;马中骐. 杨－巴克斯特方程与量子包络代数. 北京:科学出版社,1993; Chari V, Pressley A. A Guide to Quantum Groups. Cambridge University Press, London, 1993.

理含义. 在动力系统中, 量子对称性是保持运动方程协变的隐含对称性; 量子群被用来构造与求解格点模型; 共形场论提供量子群的几何解释; 而具量子对称性的光谱理论与实验很好地符合. 当然, 这些并非量子对称性的全部, 很多重要的进展这里都没有涉及, 特别是我们没有讨论 Yangian 与量子仿射对称性. 最近的工作表明, 高温超导理论中广泛研究的 Hubbard 模型具有 Yangian 对称性, 而量子仿射群的表示论是计算格点模型关联函数的有效方法. 有关这方面的研究正在迅速取得进展[①].

我国物理与数学家在量子群与量子对称性方面取得了很多有影响的成果. 老一辈科学家不但是这方面研究的骨干力量, 而且不惜余力地培养并支持青年人迅速进入研究的前沿. 形成国际上这方面研究的一支重要力量.

本书的后两章是理论物理范畴, 恰巧在笔者写此手记之际国际物理界发生了一件大事.

2021 年 7 月 24 日 (北京时间), 当代最伟大的物理学家之一史蒂芬·温伯格 (Steven Weinberg) 教授在美国逝世, 享年 88 岁. 温伯格教授在物理学的众多领域做出了开创性贡献, 特别是建立了弱相互作用和电磁相互作用的统一理论, 同时也是粒子物理标准模型的创立者之一. 温伯格教授还热衷于科学传播, 著有大量的科普图书和文章, 产生了广泛且深远的影响. 上海交通大学钮卫星教授写了一篇长文以此缅怀温伯格教授. 内容如下:

---

① Jimbo M, Miwa T. Preprint RIMS - 981(1994).

温伯格其人

1967年,温伯格提出了统一电磁作用和弱相互作用的模型,即电弱统一理论,而后被实验证实.电弱统一理论统一了电磁力和弱相互作用力,温伯格也因提出该理论与另外两位物理学家分享了1979年的诺贝尔物理学奖.

1977年,温伯格出版了他的科普名著《最初三分钟》(The First Three Minutes),通俗而精彩地介绍了关于宇宙起源的现代观念,一时洛阳纸贵,共被译成二十二种文字.

温伯格生前执教于德克萨斯大学奥斯汀分校物理与天文系.他曾在康奈尔大学、哥本哈根大学和普林斯顿大学学习,获得十多所著名大学的荣誉学位,先后任教于哥伦比亚大学、加利福尼亚大学伯克利分校、麻省理工学院和哈佛大学.他还当选为英国皇家学会会员、美国科学院院士、美国科学与艺术研究院院士、美国哲学学会会员等.1991年,布什总统在白宫授予他国家科学奖章.他还是美国物理学学会安德鲁·吉芒特奖和美国斯蒂尔基金会-美国物理学会科学作品奖的获得者.

在温伯格获得的所有奖项中尤其值得一提的一项是1999年他获得的洛克菲勒大学"刘易斯·托马斯奖"(Lewis Thomas Prize).这个奖项授予这样一些科学家作者:"他们的意见和观点能向人们揭示科学的美学和哲学维度,不仅提供新的信息,而且还能像诗歌和绘画一样引起人们的沉思,甚至启示."温伯格因"在充满热情地、清晰地传达基础物理学的观念、历史、解释力和美学维度方面所取得的杰出成就"而获得该奖.

获得"刘易斯·托马斯奖"的科学家又被叫作"诗人科学家",这个称号在肯定他们的写作技巧的同时也充分概括了他们对科学的美学和哲学层面上的思考.事实上,被用来命名这个奖项的美国微生物学家和免疫学家刘易斯·托马斯(Lewis Thomas,1913—1993)就是这样一个把科学写得跟文学作品一样优美的成功典范,他获得了用自己名字命名的这个奖.

温伯格主要是因为他为普通读者写的《最初三分钟》和《终极理论之梦》而获奖的.其实,温伯格还经常在诸如《纽约书评》《科学美国人》等一些大众传媒上发表一些阐明自己对科学的看法和观点的文章,在这些文章中,他以同样优美、生动的文笔清晰地表达了自己是一位理性论者、还原论者、实在论者和非宗教论者,有些文章还致力于在科学的文化对手们面前为科学辩护.

2001年,这些文章的一部分被编辑为一册出版.2004年底,有了中译本《仰望苍穹:科学反击文化敌手》(*Facing up:Science and its cultural adversaries*).在温伯格的这些作品中,以他对文字的娴熟驾驭和在文字中浸透的对物理学的热情来说,是无愧于"诗人科学家"的称号的.

科学的文化敌手们

按照一句西方谚语:在木匠眼里,月亮是木头做的,而在基本粒子物理学家温伯格眼里,宇宙是由基本粒子构成的,这是最理所当然的事情了.自从古希腊德谟克利特和他的老师留基伯提出世界是由不可分割的最小粒子原子构成以来,原子论一直是唯物论的中坚.

虽然原子论的内容被两千多年以后的物理学家们大大丰富了,并且原子也早已被打破,物理学家发现了更为基本的粒子,在研究更小尺度的结构——超弦,但是无论是德谟克利特的原子也好,还是现在的基本粒子也好,它们都被认为是客观的、实在的.用温伯格的话来说,"跟操场上的石头一样真实."

可是有人却说:"诸如夸克这一类基本粒子是一群科学家建构出来的,进而说科学事实、科学理论也是建构出来的,科学理论只不过是一种整理我们的经验的人为方式".是可忍,孰不可忍?温伯格肯定是觉得有义务站出来驳斥这种言论,而且他显然也是最有权利和资格站出来保卫科学的人.

温伯格捍卫科学的这些文章大多收集在《仰望苍穹》中.按照温伯格的交代,这个书名有三层意思.

第一层意思是所取正如书的封面所用的丹麦天文学家第谷雕像的姿态:翘首仰望苍天;

第二层意思是要强调正确面对和正视人类自身在宇宙中位置的必要性,同时也表明对那些宗教神创论等形形色色的科学的文化敌手要采取勇敢直面的态度;

第三层意思是要表明这种直面正视和翘首仰望的姿态与匍匐在地的祈祷者的姿态恰恰相反.

比较中译本书名和原书名,中文书名的正题概括性稍弱,而副题则突出了冲突的"火药味".

《仰望苍穹》英文书名中的文化敌手是复数的.这些被温伯格列为科学的文化敌手的,主要有宗教神创论者、宇宙智能设计论者和科学理论的社会建构论者、后现代主义者、女性主义者,部分科学哲学家和科学史家或者他们的部分理论也被温伯格列为科学的文化敌手.书中对以上各种论者的言论都进行了针锋相对的驳斥.

温伯格对宗教神创论和宇宙设计论的驳斥可谓决不留情,毫不妥协.对于宗教和科学的冲突,各种论述已经汗牛充栋;而对宗教的道德功能一般争议很少,人们大都承认宗教是有劝人向善的道德教化功能的.

古尔德曾经在《科学美国人》上撰文:"科学与宗教没有冲突,科学应对现实的存在,而宗教应对人类道德."然而,温伯格却连宗教的道德功能也加以否定.在"宇宙有设计者吗?"中他用历史上的事例证明:"宗教的道德色调从时代精神中受益要胜于时代精神从宗教中受益."并且旗帜鲜明地提出:"不论有没有宗教,好人都行善,坏人都作恶;可是要让好人作恶,那就要利用宗教.

在"直面奥布莱恩"中温伯格十分肯定地说:"科学家没有发现能显示人类在物理规律中或是在宇宙的初始条件中占有任何特殊地位的任何东西."在《最初三分钟》一书的结尾温伯格说了一句给他带来很多麻烦的话:"宇宙越显得可以理解,它就越显得没有意义."在《终极理论之梦》和《仰望苍穹》中温伯格都为这句话进行了辩解.他只不过是要强调,宇宙本身是没有什么意义的,人类不要把自身的价值观强加于宇宙.

对于有人提出科学与宗教之间应该展开建设性的对话这种主张,温伯格说:"对话是可以的,但绝不是建设性的."他的言下之意很明显:宗教在科学面前只有俯首帖耳"挨训"的份.在"宇宙有设计者吗?"中温伯格这样结束全文:"最伟大的科学成就之一,即便不是让智慧者不可能成为宗教信徒,那么至少是让他们可能不成为宗教信徒.我们不应该从这个成就上倒退."这可以看成是温伯格对待科学与宗教关系的一个鲜明态度.

温伯格反击的另一群科学的文化敌手就是社会建构论者、后现代主义者和女性主义者.对于科学与人文由来已久的冲突,温伯格有时也有一些建设性的想法,譬如《仰望苍穹》的第一篇就是"把科学作为文科的一门课程",这是他在一所文科院校毕业典礼上的演讲.他还深信"不论目前在科学家和公众的交流之间存在什么障碍,它们都不是不可逾越的",而且他自己就身体力行,进行通俗创作,不惜被讥为"浪费宝贵的时间和精力""大材小用",等等.然而,对社会建构论等一些科学的文化敌手,温伯格的反击是无情和严厉的.

获得过物理学博士学位的皮克林(Andrew Pickering)著有《建构夸克:粒子物理学的社会史》一书.在温伯格看来,皮克林在书中把高能物理学中的重点转移描写成"不过是一种时尚的改变,就像从印象派转向立体派,从长裙子转向短裙子."对于这种论调,温伯格以物理学家的身份说:"许多正在尽职的科学家发现,这种'社会建构论者'的观点与科学家们自身的经验不符合."并认为"按照这种(社会建构论)观点,科学理论除了是社会建构之外什么也不是,这在我看来是荒唐可笑的."

对于科学规律受发现它们的社会背景的影响这种论调,温伯格作为一名理论物理学家,习惯性地做起了思想实验:"如果我们认为科学规律的适应性足以受发现它们的社会背景的影响,那么出于一些可能的诱惑而迫使科学家去发现更无产阶级的或更妇女化的或美国的或宗教上的或印欧语系的或任何其他什么的他们所需要的规律.这是一条危险的道路……更有引起争议的危险."

确实,对于我们中国人来说,在刚刚抛弃了"无产阶级科学"和"资产阶级科学"的划分之后不久,应该更能认清这种社会建构论是有害无益的.在"索卡尔的恶作剧"

一文结尾,温伯格写道:"如果要从仍然包围着人类的不理性的趋势中保护我们自己,我们就务必要巩固和强化理性地认识世界的洞察力."这是在语重心长地提醒读者:理性来之不易,应该多多珍惜.

一些后现代主义者,"他们不仅怀疑科学的客观性,而且还厌恶客观性,他们欢迎某些比现代科学更热烈且更模糊的东西".对他们,温伯格似乎懒得大举讨伐,因为一些争论很容易滑入哲学的沼泽.譬如,科学家们相信他们在向客观真理迈进,但反对者说真理还没有被很好地定义.温伯格只是说就像给奶牛下定义是动物学家的事,给真理下定义是哲学家的事;而科学家就像农民遇见奶牛能认识它们一样,他们遇见真理也通常能认识它们.

在说到科学理论的客观性和实在性时,温伯格多次打那个石头的比方:"我们之所以说岩石是实在的,是因为它具有稳定性和不依赖于社会背景这些性质,而科学理论同样如此."

对于哲学,温伯格坦诚他不喜欢阅读从亚里士多德到阿奎那的大多数哲学著作.但是正如他在"伽利略的遗风"一篇中提到的:"现代科学再次遇上了它的由来已久的老对手,也就是伽利略曾经遭遇的主要对手,即哲学的偏见."所以他有时不得不也要正面接触哲学,特别是科学哲学.在《终极理论之梦》中专辟了一章"反对哲学"来谈论科学与哲学问题.

科学哲学家库恩是温伯格的朋友,在温伯格的书中多次提到了库恩,有一篇"库恩的不革命"甚至是温伯格专门针对《科学革命的结构》的书评.温伯格对库恩的"范式""不可通约"等概念进行了批评.

从科学史和他亲自参与的前沿物理学研究出发,温伯格批评库恩在一些科学的史实和事实上的把握有偏差.面对库恩的结论,即认为从一种范式向另一种范式的革命性转变中,科学家没有变得离真理更近一些,温伯格有点痛心疾首的味道.他写道:"正是这些结论使得他成为那些质疑科学知识的客观本质的哲学家、历史学家、社会学家和文化评论学家眼中的英雄,以及那些宁愿把科学理论在这方面说得与民主或者棒球无太大差别的社会结构论者眼中的英雄."

虽然库恩自己也抱怨过他自己不是"库恩主义者",但他的理论所产生的影响是种种后现代科学观的第一推动力.所以在温伯格看来,库恩哪怕没有喊反科学的口号,也是一位事实上的反科学者.

有趣的是,对于时下被一些人大谈特谈的科学方法,温伯格却有出人意料的态度.他认为根本没有什么科学方法.在"科学方法……和我们的生存之道"中他说:"许多科学家很少知道科学方法是什么的概念,这就好像大多数骑自行车的人对自行车是如何保持直立的概念知之甚少一样.在这两种情况里,如果想得过多,往往可能会摔跟头."这点对于正热衷于传播所谓的科学方法的我们也许有一定的借鉴意义.

美、还原论与终极理论之梦

在一部根据卡尔·萨根的同名小说改编的科幻电影《接触》中,女天文学家艾利执着地搜寻外星智慧生物的信号,她克服重重困难,终于获得成功,并被选定为地球人的使者乘坐按照外星文明传来的设计图纸制造的交通工具前往银河系中心与外星智慧生物会面.在途中艾利看到天体莫可名状的壮美,不知如何用语言来记录这种景象,只是感叹道:"应该派个诗人来!"

一般认为诗是记录和呈现美的恰当文体,诗人则是描绘和赞誉美的合适人选. 温伯格等人之所以被称作"诗人科学家",就是因为他们能发现和描述科学中的美.

在科学史上,有不少科学家发现并描述了科学理论中的美. 传说毕达哥拉斯发现了 2∶1,3∶2,4∶3 这几个数字比率跟最和谐的音程八度音、五度音和四度音一致,于是得到了"万物皆数"的信念,认为宇宙中普遍存在着数的和谐,譬如各天体到地球的距离便应该符合和谐的音程.

柏拉图继承了这种信念,认为自然界必定是完美的,天体必定做完美的运动. 亚里士多德进而认为天体必定是有完美的第五元素构成. 哥白尼为了恢复和保持这种古希腊的信念才做出了革命性的举动. 开普勒对行星运动规律的探索也一直在"宇宙存在先定的和谐"这种古希腊信念指导下进行. 而正如众所周知的,从哥白尼到开普勒,开启了科学迈向近代走向现代的大门.

许多现代物理学家也都谈到过美在科学中的作用. 1974 年,量子电动力学的创立者之一狄拉克到哈佛大学演讲,他对听讲的研究生说:"只需要去关注方程的美,而不要去管那些方程是什么意思." 温伯格当时也在座,他认为对学生来说这也许不是什么好建议,但他承认寻找物理学中的美贯穿着整个物理学的历史.

在《终极理论之梦》一书中他专门用一章"美丽的理论"来讲物理学中的美,认为美感是向着终极理论进步的标志. 1987 年,温伯格为牛顿的《自然哲学的数学原理》出版 300 周年做了一个"牛顿之梦"的纪念性演讲,其中他提到"如果我们是在谈论非常基本的现象,那么在某种程度上美学思想是重要的",他还强调:"我们所寻求的美是一种特殊类型的美……我们所寻求的美的理论,能给我们一种任何东西都不会使之改变的感觉."

温伯格还用物理学家惠勒的话来说明他的观点:"当我们最终认识到自然的终极规律之时,我们将会感到奇怪,为什么它们不是从一开始就那么明显呢!"

显然,在温伯格看来,也在许多其他物理学家看来,科学理论的美具有的一个最基本的属性,就是简单.在一篇阐述自己的世界观和认识论的短文《直面奥布莱恩》中,温伯格深信沿着解释自然的链条深究下去,用来解释的原理"看起来会变得更加简单也更加统一;我们采用越来越少的原理来解释越来越多的东西;我们会发现少数几个极其简单而又无比优美的普遍原理,即自然规律."在为《乔治》杂志写的一篇介绍自己如何发现电弱统一理论的文章《红色卡玛洛》中,温伯格把他的工作说成是"力图对复杂现象做出简单的解释".

温伯格十分注重这种位于解释链条最后端点的东西,认为这是物理学乃至科学的最终目的.在一篇《一位量子物理学家的深夜冥想》的文章中,温伯格写道:"如果你问到有关为什么事情是如此这般的任何问题……其解释总是会还原到更深层理论的形式.更深层并不是指在数学上意义更深远或更有用,而是指更接近我们解释的起点."他举例说:"许多矿物学和生理学是根据化学来解释的,化学是根据物理学来解释的,普通物质的物理学又是根据基本粒子的标准模型来解释的."

温伯格的这种思想其实已经被恰当地命名为"还原

论".1996年,另一位"刘易斯·托马斯奖"的获得者戴森(Freeman Dyson)把物理学上的还原论描述为"力图使物理现象的世界还原为一套有限的基本方程",他还把薛定谔和狄拉克在量子力学方面的工作称作是还原论的胜利,"化学和物理学的令人迷惑的复杂性被还原为两种形式的代数符号."

事实上,还原论在有些场合被用作贬义词,被认为是属于过时的机械论哲学而应该被抛弃.也有人把还原论称作"物理帝国主义",而温伯格也委婉地承认自己在某种程度上是一位物理帝国主义者,即认为物理学家提供了一套能解释其他所有事情的自然规律,其他科学看起来只不过是物理学的衍生物.

温伯格在不同场合多次阐述自己的还原论观点,他的《终极理论之梦》可以说是一首还原论的赞歌,其中专门有一章叫作"为还原论欢呼";在《仰望苍穹》一书中,也收录了"牛顿论、还原论和国会论证艺术""还原论的回归""大还原:20世纪物理学"等篇.

在"还原论的回归"一文中温伯格写道:"从牛顿时代到我们所处的时代,我们已经看到,我们知道如何去解释的现象的范围在不断地扩大,而用于这些解释的理论在简单性和普遍性方面也在不断地得到改善.""由于简单而普遍的定律,自然的一切就是它现有的这种方式(带有初始条件和历史偶然性的某种限制),所有其他科学定律在某种意义上都可以还原为这种定律."

按照还原论的思路,沿着解释的链条,终点就是那解释一切的终极理论.在《牛顿之梦》中温伯格说:"把那些能够解释,为什么每件事情理当如此的少数几个简单的原则系统化.这是牛顿的梦想."这当然也是温伯格的梦想,《终极理论之梦》充满激情地表达了对实现这个梦想所抱有的信心.

在《物理学与历史》一文最后,温伯格写道:"终极理论将会是一种有效性不受限制的理论,一种能适用于整个宇宙中所有现象的理论,一旦最终达到,这种理论将成为我们关于这个世界的知识的一个永恒的部分.那时我们作为基本粒子物理学家的工作将会结束."

即便有反对的声音认为人类不可能找到这样的终极理论,但温伯格直至临终还一直在坚持,仍在追求这最后的理论.温伯格的这种坚持和追求哪怕是堂吉诃德式的,也仍旧是可敬的.人类对自然的了解和适应因为这样的坚持和追求而在不断进步.

正如本文开头所引的那句 E. W. Montron 的名言所示:数学、物理不分家,所以,虽然我们工作室专做数学,但也兼顾物理,特别是理论物理、计算物理、非线性物理等领域.正如有人从几位大诗人的传记出发,归纳出一流诗人的成长定律:脾气要怪一点,身体要差一点,神经要脆弱一点一样,我们似乎也可以从物理学家的成才之路中总结出一点规律性的东西,即数学要好一点!

刘培杰
2022 年 4 月 21 日
于哈工大

刘培杰数学工作室
已出版(即将出版)图书目录——原版影印

书　名	出版时间	定　价	编号
数学物理大百科全书.第1卷(英文)	2016—01	418.00	508
数学物理大百科全书.第2卷(英文)	2016—01	408.00	509
数学物理大百科全书.第3卷(英文)	2016—01	396.00	510
数学物理大百科全书.第4卷(英文)	2016—01	408.00	511
数学物理大百科全书.第5卷(英文)	2016—01	368.00	512
zeta函数,q-zeta函数,相伴级数与积分(英文)	2015—08	88.00	513
微分形式:理论与练习(英文)	2015—08	58.00	514
离散与微分包含的逼近和优化(英文)	2015—08	58.00	515
艾伦·图灵:他的工作与影响(英文)	2016—01	98.00	560
测度理论概率导论,第2版(英文)	2016—01	88.00	561
带有潜在故障恢复系统的半马尔柯夫模型控制(英文)	2016—01	98.00	562
数学分析原理(英文)	2016—01	88.00	563
随机偏微分方程的有效动力学(英文)	2016—01	88.00	564
图的谱半径(英文)	2016—01	58.00	565
量子机器学习中数据挖掘的量子计算方法(英文)	2016—01	98.00	566
量子物理的非常规方法(英文)	2016—01	118.00	567
运输过程的统一非局部理论:广义波尔兹曼物理动力学,第2版(英文)	2016—01	198.00	568
量子力学与经典力学之间的联系在原子、分子及电动力学系统建模中的应用(英文)	2016—01	58.00	569
算术域(英文)	2018—01	158.00	821
高等数学竞赛:1962—1991年的米洛克斯·史怀哲竞赛(英文)	2018—01	128.00	822
用数学奥林匹克精神解决数论问题(英文)	2018—01	108.00	823
代数几何(德文)	2018—04	68.00	824
丢番图逼近论(英文)	2018—01	78.00	825
代数几何学基础教程(英文)	2018—01	98.00	826
解析数论入门课程(英文)	2018—01	78.00	827
数论中的丢番图问题(英文)	2018—01	78.00	829
数论(梦幻之旅):第五届中日数论研讨会演讲集(英文)	2018—01	68.00	830
数论新应用(英文)	2018—01	68.00	831
数论(英文)	2018—01	78.00	832

刘培杰数学工作室
已出版(即将出版)图书目录——原版影印

书　名	出版时间	定　价	编号
湍流十讲(英文)	2018—04	108.00	886
无穷维李代数:第3版(英文)	2018—04	98.00	887
等值、不变量和对称性(英文)	2018—04	78.00	888
解析数论(英文)	2018—09	78.00	889
《数学原理》的演化:伯特兰·罗素撰写第二版时的手稿与笔记(英文)	2018—04	108.00	890
哈密尔顿数学论文集(第4卷):几何学、分析学、天文学、概率和有限差分等(英文)	2019—05	108.00	891
偏微分方程全局吸引子的特性(英文)	2018—09	108.00	979
整函数与下调和函数(英文)	2018—09	118.00	980
幂等分析(英文)	2018—09	118.00	981
李群,离散子群与不变量理论(英文)	2018—09	108.00	982
动力系统与统计力学(英文)	2018—09	118.00	983
表示论与动力系统(英文)	2018—09	118.00	984
分析学练习.第1部分(英文)	2021—01	88.00	1247
分析学练习.第2部分,非线性分析(英文)	2021—01	88.00	1248
初级统计学:循序渐进的方法:第10版(英文)	2019—05	68.00	1067
工程师与科学家微分方程用书:第4版(英文)	2019—07	58.00	1068
大学代数与三角学(英文)	2019—06	78.00	1069
培养数学能力的途径(英文)	2019—07	38.00	1070
工程师与科学家统计学:第4版(英文)	2019—06	58.00	1071
贸易与经济中的应用统计学:第6版(英文)	2019—06	58.00	1072
傅立叶级数和边值问题:第8版(英文)	2019—05	48.00	1073
通往天文学的途径:第5版(英文)	2019—05	58.00	1074
拉马努金笔记.第1卷(英文)	2019—06	165.00	1078
拉马努金笔记.第2卷(英文)	2019—06	165.00	1079
拉马努金笔记.第3卷(英文)	2019—06	165.00	1080
拉马努金笔记.第4卷(英文)	2019—06	165.00	1081
拉马努金笔记.第5卷(英文)	2019—06	165.00	1082
拉马努金遗失笔记.第1卷(英文)	2019—06	109.00	1083
拉马努金遗失笔记.第2卷(英文)	2019—06	109.00	1084
拉马努金遗失笔记.第3卷(英文)	2019—06	109.00	1085
拉马努金遗失笔记.第4卷(英文)	2019—06	109.00	1086
数论:1976年纽约洛克菲勒大学数论会议记录(英文)	2020—06	68.00	1145
数论:卡本代尔1979:1979年在南伊利诺伊卡本代尔大学举行的数论会议记录(英文)	2020—06	78.00	1146
数论:诺德韦克豪特1983:1983年在诺德韦克豪特举行的Journees Arithmetiques数论大会会议记录(英文)	2020—06	68.00	1147
数论:1985—1988年在纽约城市大学研究生院和大学中心举办的研讨会(英文)	2020—06	68.00	1148

刘培杰数学工作室
已出版(即将出版)图书目录——原版影印

书　名	出版时间	定　价	编号
数论:1987年在乌尔姆举行的Journees Arithmetiques数论大会会议记录(英文)	2020—06	68.00	1149
数论:马德拉斯1987:1987年在马德拉斯安娜大学举行的国际拉马努金百年纪念大会会议记录(英文)	2020—06	68.00	1150
解析数论:1988年在东京举行的日法研讨会会议记录(英文)	2020—06	68.00	1151
解析数论:2002年在意大利切特拉罗举行的C.I.M.E.暑期班演讲集(英文)	2020—06	68.00	1152
量子世界中的蝴蝶:最迷人的量子分形故事(英文)	2020—06	118.00	1157
走进量子力学(英文)	2020—06	118.00	1158
计算物理学概论(英文)	2020—06	48.00	1159
物质,空间和时间的理论:量子理论(英文)	2020—10	48.00	1160
物质,空间和时间的理论:经典理论(英文)	2020—10	48.00	1161
量子场理论:解释世界的神秘背景(英文)	2020—07	38.00	1162
计算物理学概论(英文)	2020—06	48.00	1163
行星状星云(英文)	2020—10	38.00	1164
基本宇宙学:从亚里士多德的宇宙到大爆炸(英文)	2020—08	58.00	1165
数学磁流体力学(英文)	2020—07	58.00	1166
计算科学:第1卷,计算的科学(日文)	2020—07	88.00	1167
计算科学:第2卷,计算与宇宙(日文)	2020—07	88.00	1168
计算科学:第3卷,计算与物质(日文)	2020—07	88.00	1169
计算科学:第4卷,计算与生命(日文)	2020—07	88.00	1170
计算科学:第5卷,计算与地球环境(日文)	2020—07	88.00	1171
计算科学:第6卷,计算与社会(日文)	2020—07	88.00	1172
计算科学.别卷,超级计算机(日文)	2020—07	88.00	1173
多复变函数论(日文)	2022—06	78.00	1518
复变函数入门(日文)	2022—06	78.00	1523
代数与数论:综合方法(英文)	2020—10	78.00	1185
复分析:现代函数理论第一课(英文)	2020—10	58.00	1186
斐波那契数列和卡特兰数:导论(英文)	2020—10	68.00	1187
组合推理:计数艺术介绍(英文)	2020—07	88.00	1188
二次互反律的傅里叶分析证明(英文)	2020—07	48.00	1189
旋瓦兹分布的希尔伯特变换与应用(英文)	2020—07	58.00	1190
泛函分析:巴拿赫空间理论入门(英文)	2020—07	48.00	1191
卡塔兰数入门(英文)	2019—05	68.00	1060
测度与积分(英文)	2019—04	68.00	1059
组合学手册.第一卷(英文)	2020—06	128.00	1153
-代数、局部紧群和巴拿赫-代数丛的表示.第一卷,群和代数的基本表示理论(英文)	2020—05	148.00	1154
电磁理论(英文)	2020—08	48.00	1193
连续介质力学中的非线性问题(英文)	2020—09	78.00	1195
多变量数学入门(英文)	2021—05	68.00	1317
偏微分方程入门(英文)	2021—05	88.00	1318
若尔当典范性:理论与实践(英文)	2021—07	68.00	1366
伽罗瓦理论.第4版(英文)	2021—08	88.00	1408

刘培杰数学工作室
已出版(即将出版)图书目录——原版影印

书　名	出版时间	定价	编号
典型群,错排与素数(英文)	2020－11	58.00	1204
李代数的表示:通过 gln 进行介绍(英文)	2020－10	38.00	1205
实分析演讲集(英文)	2020－10	38.00	1206
现代分析及其应用的课程(英文)	2020－10	58.00	1207
运动中的抛射物数学(英文)	2020－10	38.00	1208
2－纽结与它们的群(英文)	2020－10	38.00	1209
概率,策略和选择:博弈与选举中的数学(英文)	2020－11	58.00	1210
分析学引论(英文)	2020－11	58.00	1211
量子群:通往流代数的路径(英文)	2020－11	38.00	1212
集合论入门(英文)	2020－10	48.00	1213
酉反射群(英文)	2020－11	58.00	1214
探索数学:吸引人的证明方式(英文)	2020－11	58.00	1215
微分拓扑短期课程(英文)	2020－10	48.00	1216
抽象凸分析(英文)	2020－11	68.00	1222
费马大定理笔记(英文)	2021－03	48.00	1223
高斯与雅可比和(英文)	2021－03	78.00	1224
π与算术几何平均:关于解析数论和计算复杂性的研究(英文)	2021－01	58.00	1225
复分析入门(英文)	2021－03	48.00	1226
爱德华·卢卡斯与素性测定(英文)	2021－03	78.00	1227
通往凸分析及其应用的简单路径(英文)	2021－01	68.00	1229
微分几何的各个方面.第一卷(英文)	2021－01	58.00	1230
微分几何的各个方面.第二卷(英文)	2020－12	58.00	1231
微分几何的各个方面.第三卷(英文)	2020－12	58.00	1232
沃克流形几何学(英文)	2020－11	58.00	1233
彷射和韦尔几何应用(英文)	2020－12	58.00	1234
双曲几何学的旋转向量空间方法(英文)	2021－02	58.00	1235
积分:分析学的关键(英文)	2020－12	48.00	1236
为有天分的新生准备的分析学基础教材(英文)	2020－11	48.00	1237
数学不等式.第一卷.对称多项式不等式(英文)	2021－03	108.00	1273
数学不等式.第二卷.对称有理不等式与对称无理不等式(英文)	2021－03	108.00	1274
数学不等式.第三卷.循环不等式与非循环不等式(英文)	2021－03	108.00	1275
数学不等式.第四卷.Jensen不等式的扩展与加细(英文)	2021－03	108.00	1276
数学不等式.第五卷.创建不等式与解不等式的其他方法(英文)	2021－04	108.00	1277

刘培杰数学工作室
已出版(即将出版)图书目录——原版影印

书　　名	出版时间	定　价	编号
冯·诺伊曼代数中的谱位移函数:半有限冯·诺伊曼代数中的谱位移函数与谱流(英文)	2021－06	98.00	1308
链接结构:关于嵌入完全图的直线中链接单形的组合结构(英文)	2021－05	58.00	1309
代数几何方法.第1卷(英文)	2021－06	68.00	1310
代数几何方法.第2卷(英文)	2021－06	68.00	1311
代数几何方法.第3卷(英文)	2021－06	58.00	1312
代数、生物信息和机器人技术的算法问题.第四卷,独立恒等式系统(俄文)	2020－08	118.00	1199
代数、生物信息和机器人技术的算法问题.第五卷,相对覆盖性和独立可拆分恒等式系统(俄文)	2020－08	118.00	1200
代数、生物信息和机器人技术的算法问题.第六卷,恒等式和准恒等式的相等 问题、可推导性和可实现性(俄文)	2020－08	128.00	1201
分数阶微积分的应用:非局部动态过程,分数阶导热系数(俄文)	2021－01	68.00	1241
泛函分析问题与练习:第2版(俄文)	2021－01	98.00	1242
集合论、数学逻辑和算法论问题:第5版(俄文)	2021－01	98.00	1243
微分几何和拓扑短期课程(俄文)	2021－01	98.00	1244
素数规律(俄文)	2021－01	88.00	1245
无穷边值问题解的递减:无界域中的拟线性椭圆和抛物方程(俄文)	2021－01	48.00	1246
微分几何讲义(俄文)	2020－12	98.00	1253
二次型和矩阵(俄文)	2021－01	98.00	1255
积分和级数.第2卷,特殊函数(俄文)	2021－01	168.00	1258
积分和级数.第3卷,特殊函数补充:第2版(俄文)	2021－01	178.00	1264
几何图上的微分方程(俄文)	2021－01	138.00	1259
数论教程:第2版(俄文)	2021－01	98.00	1260
非阿基米德分析及其应用(俄文)	2021－03	98.00	1261
古典群和量子群的压缩(俄文)	2021－03	98.00	1263
数学分析习题集.第3卷,多元函数:第3版(俄文)	2021－03	98.00	1266
数学习题:乌拉尔国立大学数学力学系大学生奥林匹克(俄文)	2021－03	98.00	1267
柯西定理和微分方程的特解(俄文)	2021－03	98.00	1268
组合极值问题及其应用:第3版(俄文)	2021－03	98.00	1269
数学词典(俄文)	2021－01	98.00	1271
确定性混沌分析模型(俄文)	2021－06	168.00	1307
精选初等数学习题和定理.立体几何.第3版(俄文)	2021－03	68.00	1316
微分几何习题:第3版(俄文)	2021－05	98.00	1336
精选初等数学习题和定理.平面几何.第4版(俄文)	2021－05	68.00	1335
曲面理论在欧氏空间 E_n 中的直接表示(俄文)	2022－01	68.00	1444
维纳－霍普夫离散算子和托普利兹算子:某些可数赋范空间中的诺特性和可逆性(俄文)	2022－03	108.00	1496
Maple中的数论:数论中的计算机计算(俄文)	2022－03	88.00	1497
贝尔曼和克努特问题及其概括:加法运算的复杂性(俄文)	2022－03	138.00	1498

V

刘培杰数学工作室
已出版(即将出版)图书目录——原版影印

书　名	出版时间	定　价	编号
复分析:共形映射(俄文)	2022-07	48.00	1542
微积分代数样条和多项式及其在数值方法中的应用(俄文)	2022-08	128.00	1543
蒙特卡罗方法中的随机过程和场模型:算法和应用(俄文)	2022-08	88.00	1544
狭义相对论与广义相对论:时空与引力导论(英文)	2021-07	88.00	1319
束流物理学和粒子加速器的实践介绍:第2版(英文)	2021-07	88.00	1320
凝聚态物理中的拓扑和微分几何简介(英文)	2021-05	88.00	1321
混沌映射:动力学、分形学和快速涨落(英文)	2021-05	128.00	1322
广义相对论:黑洞、引力波和宇宙学介绍(英文)	2021-06	68.00	1323
现代分析电磁均质化(英文)	2021-06	68.00	1324
为科学家提供的基本流体动力学(英文)	2021-06	88.00	1325
视觉天文学:理解夜空的指南(英文)	2021-06	68.00	1326
物理学中的计算方法(英文)	2021-06	68.00	1327
单星的结构与演化:导论(英文)	2021-06	108.00	1328
超越居里:1903年至1963年物理界四位女性及其著名发现(英文)	2021-06	68.00	1329
范德瓦尔斯流体热力学的进展(英文)	2021-06	68.00	1330
先进的托卡马克稳定性理论(英文)	2021-06	88.00	1331
经典场论导论:基本相互作用的过程(英文)	2021-07	88.00	1332
光致电离量子动力学方法原理	2021-07	108.00	1333
经典域论和应力:能量张量(英文)	2021-05	88.00	1334
非线性太赫兹光谱的概念与应用(英文)	2021-06	68.00	1337
电磁学中的无穷空间并矢格林函数(英文)	2021-06	88.00	1338
物理科学基础数学.第1卷,齐次边值问题、傅里叶方法和特殊函数(英文)	2021-07	108.00	1339
离散量子力学(英文)	2021-07	68.00	1340
核磁共振的物理学和数学(英文)	2021-07	108.00	1341
分子水平的静电学(英文)	2021-08	68.00	1342
非线性波:理论、计算机模拟、实验(英文)	2021-06	108.00	1343
石墨烯光学:经典问题的电解解决方案(英文)	2021-06	68.00	1344
超材料多元宇宙(英文)	2021-07	68.00	1345
银河系外的天体物理学(英文)	2021-07	68.00	1346
原子物理学(英文)	2021-07	68.00	1347
将光打结:将拓扑学应用于光学(英文)	2021-07	68.00	1348
电磁学:问题与解法(英文)	2021-07	88.00	1364
海浪的原理:介绍量子力学的技巧与应用(英文)	2021-07	108.00	1365
多孔介质中的流体:输运与相变(英文)	2021-07	68.00	1372
洛伦兹群的物理学(英文)	2021-08	68.00	1373
物理导论的数学方法和解决问题手册(英文)	2021-08	68.00	1374
非线性波数学物理学入门(英文)	2021-08	88.00	1376
波:基本原理和动力学(英文)	2021-07	68.00	1377
光电子量子计量学.第1卷,基础(英文)	2021-07	88.00	1383
光电子量子计量学.第2卷,应用与进展(英文)	2021-07	88.00	1384
复杂流的格子玻尔兹曼建模的工程应用(英文)	2021-08	68.00	1393
电偶极矩挑战(英文)	2021-08	108.00	1394
电动力学:问题与解法(英文)	2021-09	68.00	1395
自由电子激光的经典理论(英文)	2021-08	68.00	1397

刘培杰数学工作室
已出版(即将出版)图书目录——原版影印

书　名	出版时间	定　价	编号
曼哈顿计划——核武器物理学简介(英文)	2021-09	68.00	1401
粒子物理学(英文)	2021-09	68.00	1402
引力场中的量子信息(英文)	2021-09	128.00	1403
器件物理学的基本经典力学(英文)	2021-09	68.00	1404
等离子体物理及其空间应用导论.第1卷,基本原理和初步过程(英文)	2021-09	68.00	1405
拓扑与超弦理论焦点问题(英文)	2021-07	58.00	1349
应用数学:理论、方法与实践(英文)	2021-07	78.00	1350
非线性特征值问题:牛顿型方法与非线性瑞利函数(英文)	2021-07	58.00	1351
广义膨胀和齐性:利用齐性构造齐次系统的李雅普诺夫函数和控制律(英文)	2021-06	48.00	1352
解析数论焦点问题(英文)	2021-07	58.00	1353
随机微分方程:动态系统方法(英文)	2021-07	58.00	1354
经典力学与微分几何(英文)	2021-07	58.00	1355
负定相交形式流形上的瞬子模空间几何(英文)	2021-07	68.00	1356
广义卡塔兰轨道分析:广义卡塔兰轨道计算数字的方法(英文)	2021-07	48.00	1367
洛伦兹方法的变分:二维与三维洛伦兹方法(英文)	2021-08	38.00	1378
几何、分析和数论精编(英文)	2021-08	68.00	1380
从一个新角度看数论:通过遗传方法引入现实的概念(英文)	2021-07	58.00	1387
动力系统:短期课程(英文)	2021-08	68.00	1382
几何路径:理论与实践(英文)	2021-08	48.00	1385
论天体力学中某些问题的不可积性(英文)	2021-07	88.00	1396
广义斐波那契数列及其性质(英文)	2021-08	38.00	1386
对称函数和麦克唐纳多项式:余代数结构与 Kawanaka 恒等式(英文)	2021-09	38.00	1400
杰弗里·英格拉姆·泰勒科学论文集:第1卷.固体力学(英文)	2021-05	78.00	1360
杰弗里·英格拉姆·泰勒科学论文集:第2卷.气象学、海洋学和湍流(英文)	2021-05	68.00	1361
杰弗里·英格拉姆·泰勒科学论文集:第3卷.空气动力学以及落弹数和爆炸的力学(英文)	2021-05	68.00	1362
杰弗里·英格拉姆·泰勒科学论文集:第4卷.有关流体力学(英文)	2021-05	58.00	1363

刘培杰数学工作室
已出版(即将出版)图书目录——原版影印

书　名	出版时间	定　价	编号
非局域泛函演化方程:积分与分数阶(英文)	2021-08	48.00	1390
理论工作者的高等微分几何:纤维丛、射流流形和拉格朗日理论(英文)	2021-08	68.00	1391
半线性退化椭圆微分方程:局部定理与整体定理(英文)	2021-07	48.00	1392
非交换几何、规范理论和重整化:一般简介与非交换量子场论的重整化(英文)	2021-09	78.00	1406
数论论文集:拉普拉斯变换和带有数论系数的幂级数(俄文)	2021-09	48.00	1407
挠理论专题:相对极大值,单射与扩充模(英文)	2021-09	88.00	1410
强正则图与欧几里得若尔当代数:非通常关系中的启示(英文)	2021-10	48.00	1411
拉格朗日几何和哈密顿几何:力学的应用(英文)	2021-10	48.00	1412
时滞微分方程与差分方程的振动理论:二阶与三阶(英文)	2021-10	98.00	1417
卷积结构与几何函数理论:用以研究特定几何函数理论方向的分数阶微积分算子与卷积结构(英文)	2021-10	48.00	1418
经典数学物理的历史发展(英文)	2021-10	78.00	1419
扩展线性丢番图问题(英文)	2021-10	38.00	1420
一类混沌动力系统的分歧分析与控制:分歧分析与控制(英文)	2021-11	38.00	1421
伽利略空间和伪伽利略空间中一些特殊曲线的几何性质(英文)	2022-01	68.00	1422
一阶偏微分方程:哈密尔顿—雅可比理论(英文)	2021-11	48.00	1424
各向异性黎曼多面体的反问题:分段光滑的各向异性黎曼多面体反边界谱问题:唯一性(英文)	2021-11	38.00	1425
项目反应理论手册.第一卷,模型(英文)	2021-11	138.00	1431
项目反应理论手册.第二卷,统计工具(英文)	2021-11	118.00	1432
项目反应理论手册.第三卷,应用(英文)	2021-11	138.00	1433
二次无理数:经典数论入门(英文)	2022-05	138.00	1434
数,形与对称性:数论,几何和群论导论(英文)	2022-05	128.00	1435
有限域手册(英文)	2021-11	178.00	1436
计算数论(英文)	2021-11	148.00	1437
拟群与其表示简介(英文)	2021-11	88.00	1438
数论与密码学导论:第二版(英文)	2022-01	148.00	1423

刘培杰数学工作室
已出版（即将出版）图书目录——原版影印

书　　名	出版时间	定　价	编号
几何分析中的柯西变换与黎兹变换:解析调和容量和李普希兹调和容量、变化和振荡以及一致可求长性(英文)	2021—12	38.00	1465
近似不动点定理及其应用(英文)	2022—05	28.00	1466
局部域的相关内容解析:对局部域的扩展及其伽罗瓦群的研究(英文)	2022—01	38.00	1467
反问题的二进制恢复方法(英文)	2022—03	28.00	1468
对几何函数中某些类的各个方面的研究:复变量理论(英文)	2022—01	38.00	1469
覆盖、对应和非交换几何(英文)	2022—01	28.00	1470
最优控制理论中的随机线性调节器问题:随机最优线性调节器问题(英文)	2022—01	38.00	1473
正交分解法:涡流流体动力学应用的正交分解法(英文)	2022—01	38.00	1475
芬斯勒几何的某些问题(英文)	2022—03	38.00	1476
受限三体问题(英文)	2022—05	38.00	1477
利用马利亚万微积分进行 Greeks 的计算:连续过程、跳跃过程中的马利亚万微积分和金融领域中的 Greeks(英文)	2022—05	48.00	1478
经典分析和泛函分析的应用:分析学的应用(英文)	2022—03	38.00	1479
特殊芬斯勒空间的探究(英文)	2022—03	48.00	1480
某些图形的施泰纳距离的细谷多项式:细谷多项式与图的维纳指数(英文)	2022—05	38.00	1481
图论问题的遗传算法:在新鲜与模糊的环境中(英文)	2022—05	48.00	1482
多项式映射的渐近簇(英文)	2022—05	38.00	1483
一维系统中的混沌:符号动力学,映射序列,一致收敛和沙可夫斯基定理(英文)	2022—05	38.00	1509
多维边界层流动与传热分析:粘性流体流动的数学建模与分析(英文)	2022—05	38.00	1510
演绎理论物理学的原理:一种基于量子力学波函数的逐次置信估计的一般理论的提议(英文)	2022—05	38.00	1511
R^2 和 R^3 中的仿射弹性曲线:概念和方法(英文)	2022—08	38.00	1512
算术数列中除数函数的分布:基本内容、调查、方法、第二矩、新结果(英文)	2022—05	28.00	1513
抛物型狄拉克算子和薛定谔方程:不定常薛定谔方程的抛物型狄拉克算子及其应用(英文)	2022—07	28.00	1514
黎曼-希尔伯特问题与量子场论:可积重正化、戴森-施温格方程(英文)	2022—08	38.00	1515
代数结构和几何结构的形变理论(英文)	2022—08	48.00	1516
概率结构和模糊结构上的不动点:概率结构和直觉模糊度量空间的不动点定理(英文)	2022—08	38.00	1517

刘培杰数学工作室
已出版(即将出版)图书目录——原版影印

书　名	出版时间	定　价	编号
反若尔当对:简单反若尔当对的自同构	2022—07	28.00	1533
对某些黎曼—芬斯勒空间变换的研究:芬斯勒几何中的某些变换	2022—07	38.00	1534
内诣零流形映射的尼尔森数的阿诺索夫关系	即将出版		1535
与广义积分变换有关的分数次演算:对分数次演算的研究	即将出版		1536
强子的芬斯勒几何和吕拉几何(宇宙学方面):强子结构的芬斯勒几何和吕拉几何(拓扑缺陷)	即将出版		1537
一种基于混沌的非线性最优化问题:作业调度问题	即将出版		1538
广义概率论发展前景:关于趣味数学与置信函数实际应用的一些原创观点	即将出版		1539
纽结与物理学:第二版(英文)	2022—09	118.00	1547
正交多项式和q—级数的前沿(英文)	即将出版		1548
算子理论问题集(英文)	即将出版		1549
抽象代数:群、环与域的应用导论:第二版(英文)	即将出版		1550
菲尔兹奖得主演讲集:第三版(英文)	即将出版		1551
多元实函数教程(英文)	即将出版		1552

联系地址:哈尔滨市南岗区复华四道街 10 号　哈尔滨工业大学出版社刘培杰数学工作室
网　　址:http://lpj.hit.edu.cn/
邮　　编:150006
联系电话:0451—86281378　　13904613167
E-mail:lpj1378@163.com